# CAMBRIDGE LIBRARY COLLECTION

*Books of enduring scholarly value*

## Botany and Horticulture

Until the nineteenth century, the investigation of natural phenomena, plants and animals was considered either the preserve of elite scholars or a pastime for the leisured upper classes. As increasing academic rigour and systematisation was brought to the study of 'natural history', its subdisciplines were adopted into university curricula, and learned societies (such as the Royal Horticultural Society, founded in 1804) were established to support research in these areas. A related development was strong enthusiasm for exotic garden plants, which resulted in plant collecting expeditions to every corner of the globe, sometimes with tragic consequences. This series includes accounts of some of those expeditions, detailed reference works on the flora of different regions, and practical advice for amateur and professional gardeners.

## Botany for Ladies

Jane Loudon (1807–58), the Mrs Beeton of the Victorian gardening world, wrote several popular books on horticulture and botany specifically for women. Her enthusiasm for plants and gardening was encouraged by her husband, the landscape designer John Claudius Loudon, whom she married in 1830. Her *Instructions in Gardening for Ladies* (also reissued in this series) was enormously successful, and she followed it up in 1842 with this volume on botany, in which she uses the natural system of classification. The 'grand object' of the work is 'to enable my readers to find out the name of a plant when they see it ... or, if they hear or read the name ... to make that name intelligible to them'. She takes her readers through the botanical orders, using a familiar plant as an exemplar for each, and then presents de Candolle's systematic description of plant species.

Cambridge University Press has long been a pioneer in the reissuing of out-of-print titles from its own backlist, producing digital reprints of books that are still sought after by scholars and students but could not be reprinted economically using traditional technology. The Cambridge Library Collection extends this activity to a wider range of books which are still of importance to researchers and professionals, either for the source material they contain, or as landmarks in the history of their academic discipline.

Drawing from the world-renowned collections in the Cambridge University Library and other partner libraries, and guided by the advice of experts in each subject area, Cambridge University Press is using state-of-the-art scanning machines in its own Printing House to capture the content of each book selected for inclusion. The files are processed to give a consistently clear, crisp image, and the books finished to the high quality standard for which the Press is recognised around the world. The latest print-on-demand technology ensures that the books will remain available indefinitely, and that orders for single or multiple copies can quickly be supplied.

The Cambridge Library Collection brings back to life books of enduring scholarly value (including out-of-copyright works originally issued by other publishers) across a wide range of disciplines in the humanities and social sciences and in science and technology.

# Botany for Ladies

*Or, A Popular Introduction
to the Natural System of Plants,
According to the Classification of De Candolle*

JANE LOUDON

# CAMBRIDGE
## UNIVERSITY PRESS

University Printing House, Cambridge, CB2 8BS, United Kingdom

Cambridge University Press is part of the University of Cambridge.

It furthers the University's mission by disseminating knowledge in the pursuit of education, learning and research at the highest international levels of excellence.

www.cambridge.org
Information on this title: www.cambridge.org/9781108075633

© in this compilation Cambridge University Press 2015

This edition first published 1842
This digitally printed version 2015

ISBN 978-1-108-07563-3 Paperback

# Selected botanical reference works available in the
# CAMBRIDGE LIBRARY COLLECTION

al-Shirazi, Noureddeen Mohammed Abdullah (compiler), translated by Francis Gladwin: *Ulfáz Udwiyeh, or the Materia Medica* (1793) [ISBN 9781108056090]

Arber, Agnes: *Herbals: Their Origin and Evolution* (1938) [ISBN 9781108016711]

Arber, Agnes: *Monocotyledons* (1925) [ISBN 9781108013208]

Arber, Agnes: *The Gramineae* (1934) [ISBN 9781108017312]

Arber, Agnes: *Water Plants* (1920) [ISBN 9781108017329]

Bower, F.O.: *The Ferns (Filicales)* (3 vols., 1923–8) [ISBN 9781108013192]

Candolle, Augustin Pyramus de, and Sprengel, Kurt: *Elements of the Philosophy of Plants* (1821) [ISBN 9781108037464]

Cheeseman, Thomas Frederick: *Manual of the New Zealand Flora* (2 vols., 1906) [ISBN 9781108037525]

Cockayne, Leonard: *The Vegetation of New Zealand* (1928) [ISBN 9781108032384]

Cunningham, Robert O.: *Notes on the Natural History of the Strait of Magellan and West Coast of Patagonia* (1871) [ISBN 9781108041850]

Gwynne-Vaughan, Helen: *Fungi* (1922) [ISBN 9781108013215]

Henslow, John Stevens: *A Catalogue of British Plants Arranged According to the Natural System* (1829) [ISBN 9781108061728]

Henslow, John Stevens: *A Dictionary of Botanical Terms* (1856) [ISBN 9781108001311]

Henslow, John Stevens: *Flora of Suffolk* (1860) [ISBN 9781108055673]

Henslow, John Stevens: *The Principles of Descriptive and Physiological Botany* (1835) [ISBN 9781108001861]

Hogg, Robert: *The British Pomology* (1851) [ISBN 9781108039444]

Hooker, Joseph Dalton, and Thomson, Thomas: *Flora Indica* (1855) [ISBN 9781108037495]

Hooker, Joseph Dalton: *Handbook of the New Zealand Flora* (2 vols., 1864–7)
[ISBN 9781108030410]

Hooker, William Jackson: *Icones Plantarum* (10 vols., 1837–54)
[ISBN 9781108039314]

Hooker, William Jackson: *Kew Gardens* (1858) [ISBN 9781108065450]

Jussieu, Adrien de, edited by J.H. Wilson: *The Elements of Botany* (1849)
[ISBN 9781108037310]

Lindley, John: *Flora Medica* (1838) [ISBN 9781108038454]

Müller, Ferdinand von, edited by William Woolls: *Plants of New South Wales*
(1885) [ISBN 9781108021050]

Oliver, Daniel: *First Book of Indian Botany* (1869) [ISBN 9781108055628]

Pearson, H.H.W., edited by A.C. Seward: *Gnetales* (1929)
[ISBN 9781108013987]

Perring, Franklyn Hugh et al.: *A Flora of Cambridgeshire* (1964)
[ISBN 9781108002400]

Sachs, Julius, edited and translated by Alfred Bennett, assisted by W.T. Thiselton
Dyer: *A Text-Book of Botany* (1875) [ISBN 9781108038324]

Seward, A.C.: *Fossil Plants* (4 vols., 1898–1919) [ISBN 9781108015998]

Tansley, A.G.: *Types of British Vegetation* (1911) [ISBN 9781108045063]

Traill, Catherine Parr Strickland, illustrated by Agnes FitzGibbon Chamberlin:
*Studies of Plant Life in Canada* (1885) [ISBN 9781108033756]

Tristram, Henry Baker: *The Fauna and Flora of Palestine* (1884)
[ISBN 9781108042048]

Vogel, Theodore, edited by William Jackson Hooker: *Niger Flora* (1849)
[ISBN 9781108030380]

West, G.S.: *Algae* (1916) [ISBN 9781108013222]

Woods, Joseph: *The Tourist's Flora* (1850) [ISBN 9781108062466]

For a complete list of titles in the Cambridge Library Collection please visit:
**www.cambridge.org/features/CambridgeLibraryCollection/books.htm**

# BOTANY FOR LADIES;

OR,

## A POPULAR INTRODUCTION

TO THE

## 𝔑atural 𝔖ystem of 𝔓lants,

ACCORDING TO THE CLASSIFICATION OF DE CANDOLLE.

———◆———

BY

## MRS. LOUDON,

Author of "Instructions in Gardening for Ladies," "Year-Book of
Natural History," &c. &c.

LONDON:
JOHN MURRAY, ALBEMARLE STREET.
MDCCCXLII.

# PREFACE.

When I was a child, I never could learn Botany. There was something in the Linnean system (the only one then taught) excessively repugnant to me; I never could remember the different classes and orders, and after several attempts the study was given up as one too difficult for me to master. When I married, however, I soon found the necessity of knowing something of Botany, as well as of Gardening. I always accompanied my husband in his visits to different gardens; and when we saw beautiful flowers, I was continually asking the names, though alas! these names, when I heard them, conveyed no ideas to my mind, and I was not any wiser than before. Still the natural wish to know something of what we admire, impelled me to repeat my fruitless questions; till at last, vexed at my ignorance, and ashamed of not

being able to answer the appeals which gardeners often made to me in doubtful cases,
(supposing that Mr. Loudon's wife must know
everything about plants,) I determined to learn
Botany if possible; and as my old repugnance
remained to the Linnean system, I resolved to
study the Natural one. Accordingly I began;
but when I heard that plants were divided into
the two great classes, the Vasculares and the
Cellulares, and again into the Dicotyledons or
Exogens, the Monocotyledons or Endogens,
and the Acotyledons or Acrogens, and that the
Dicotyledons were redivided into the Dichlamydeæ and Monochlamydeæ, and again into
three sub-classes, Thalamifloræ, Calycifloræ,
and Corollifloræ, I was in despair, for I thought
it quite impossible that I ever could remember
all the hard names that seemed to stand on the
very threshold of the science, as if to forbid the
entrance of any but the initiated.

Some time afterwards, as I was walking
through the gardens of the Horticultural Society at Chiswick, my attention was attracted
by a mass of the beautiful crimson flowers of

Malope grandiflora. I had never seen the plant before, and I eagerly asked the name. " It is some Malvaceous plant," answered Mr. Loudon, carelessly ; and immediately afterwards he left me to look at some trees which he was about to have drawn for his Arboretum Britannicum. " Some Malvaceous plant," thought I, as I continued looking at the splendid bed before me ; and then I remembered how much the form of these beautiful flowers resembled that of the flowers of the crimson Mallow, the botanical name of which I recollected was Malva. " I wish I could find out some other Malvaceous plant," I thought to myself; and when we soon afterwards walked through the hothouses, I continued to ask if the Chinese Hibiscus, which I saw in flower there, did not belong to Malvaceæ. I was answered in the affirmative; and I was so pleased with my newly-acquired knowledge, that I was not satisfied till I had discovered every Malvaceous plant that was in flower in the garden. I next learned to know the Cruciferous and Umbelliferous plants ; and thus I acquired a general knowledge of three extensive orders with very little trouble to myself. My attention was more fairly aroused, and by learning one order after another, I soon attained a sufficient knowledge of Botany to answer all the purposes for which I wished to learn it, without recurring to the hard words which had so much alarmed me at the outset. One great obstacle to my advancement was the difficulty I had in understanding botanical works. With the exception of Dr. Lindley's Ladies' Botany, they were all sealed books to me ; and even that did not tell half I wanted to know, though it contained a great deal I could not understand. It is so difficult for men whose knowledge has grown with their growth, and strengthened with their strength, to imagine the state of profound ignorance in which a beginner is, that even their elementary books are like the old Eton Grammar when it was written in Latin—they require a master to explain them. It is the want that I have felt that has induced me to write the following pages; in which I have endeavoured to meet the wants of those who may be now in the same difficulties that I was in myself.

The course I pursued is also that which
I shall point out to my readers. I shall first
endeavour to explain to them as clearly as I
can the botanical characteristics of the orders
which contain plants commonly grown in British gardens; and at the end of my work I shall
lay before them a slight outline of all the orders
scientifically arranged, which they may study or
not as they like. Most ladies will, however,
probably be satisfied with knowing the orders
containing popular plants; and these, I am con-
fident, they will never repent having studied.
Indeed, I do not think that I could form a
kinder wish for them, than to hope that they
may find as much pleasure in the pursuit as I
have derived from it myself. Whenever I go
into any country I have formerly visited, I feel
as though I were endowed with a new sense.
Even the very banks by the sides of the roads,
which I before thought dull and uninteresting,
now appear fraught with beauty. A new charm
seems thrown over the face of nature, and a
degree of interest is given to even the commonest weeds. I have often heard that know-

ledge is power, and I am quite sure that it
contributes greatly to enjoyment.  A man
knowing nothing of natural history, and of
course not caring for anything relating to it,
may travel from one extremity of a country to
the other, without finding anything to interest,
or even amuse him ; but the man of science,
and particularly the Botanist, cannot walk a
dozen yards along a beaten turnpike-road without finding something to excite his attention.
A wild plant in a hedge, a tuft of moss on a
wall, and even the Lichens which discolour the
stones, all present objects of interest, and of
admiration for that Almighty Power whose
care has provided the flower to shelter the
infant germ, and has laid up a stock of nourishment in the seed to supply the first wants of the
tender plant.  It has been often said that the
study of nature has a tendency to elevate and
ameliorate the mind ; and there is perhaps no
branch of Natural History which more fully
illustrates the truth of this remark than Botany.

# CONTENTS.

## PART I.

## CHAPTER I.

## CHAPTER II.

## CHAPTER III.

## CHAPTER IV.

## CHAPTER V.

## CHAPTER VI.

## CHAPTER VII.

# CHAPTER XI.

# CHAPTER XII.

# PART II.

## CHAPTER I.

PHANEROGAMOUS PLANTS—DICOTYLEDONEÆ—I. DICHLAMYDEÆ . 239

## CHAPTER II.

PHANEROGAMOUS PLANTS—DICOTYLEDONEÆ—II. MONOCHLAMYDEÆ 419

# CHAPTER III.

# CHAPTER IV.

# MODERN BOTANY FOR LADIES.

## PART I.

### INTRODUCTION.

THE following pages are intended to enable
my readers to acquire a knowledge of Botany
with as little trouble to themselves as possible.

As, however, Botany is a "wide word," I must
here premise that I only propose to treat of that
part of the science which relates to the classification of plants, according to the natural system
of Jussieu, as improved by the late Professor
De Candolle ; and that the grand object I have
in view is to enable my readers to find out the
name of a plant when they see it for the first
time ; or, if they hear or read the name of a
plant, to make that name intelligible to them.
Nothing is more natural than to ask the name of
every pretty flower we see ; but unless the in-
quirer knows something of botany, the name, if
it be a scientific one, will seem only a collection

B

of barbarous sounds, and will convey no ideas
to the mind.   Half the interest of new greenhouse plants is thus destroyed, as few of them
have English names, and strangers will soon
cease to make any inquiries respecting them
when they find they can obtain no answers that
they can understand.   Now, a very slight knowledge of botany will take away this mortifying
feeling ; and the name of a new plant, and the
ascertaining the order to which it belongs, will
recall a variety of recollections that will open up
a new source of interest and enjoyment even in
such interesting and enjoyable things as flowers
—for we never can enjoy thoroughly anything
that we do not understand.

It now only remains for me to say why I have
divided my work into two parts.   My reason is
my belief that a student will always remember
more easily a few strongly marked divisions than
a number of smaller ones, the differences between
which are only faintly perceptible.   In a more
advanced state of knowledge, it is delightful to
trace the minute shades of difference by which
the numerous orders are united, so as to form
one great whole ; but these gentle gradations
confuse a beginner.   On this account I have
thought it best to devote the first part of my
work to a few of the more important orders,
which differ most widely from each other, and

which I have described at a greater length than
my space will allow me to bestow upon the whole;
and in the second part of my work, I shall give a
short account of the whole natural system, in-
troducing the orders described in the first part,
in their proper places, so that my readers may
see how they are connected with the others.

## MISCELLANEOUS ORDERS.

### PRELIMINARY OBSERVATIONS.

In this first part I shall endeavour to fami-
liarise my readers with botanical details, as all
the orders I shall describe contain a great number of genera ; and to begin at the beginning,
I must first tell them what is here meant by
an order, and what by a genus of plants. A
genus then may be compared to a family of
children, all the plants in it being known by one
common or generic name, in addition to their
particular or specific one. Thus, if Rosa alba
be spoken of, *Rosa* is the generic name which is
common to all roses, but *alba* is the specific name
which is only applied to the white rose.

An order includes many genera, and bears
the same affinity to a nation as a genus does to

a family.  In many cases the resemblance which
the plants in each order bear to each other is
sufficiently strong to enable the student to recognise them at first sight ; in the same manner
as you may generally know a Frenchman or a
German from an Englishman, even before you
hear him speak.  But unfortunately this general
outward resemblance does not always exist, and
it is necessary for the student to become acquainted with the general construction of flowers
before the points of resemblance which have
occasioned certain genera to be linked together
to form orders, can be understood.

It is thus evident that the first step towards
a knowledge of systematic botany is to study
flowers thoroughly, and few objects of study
can be more interesting, whether we regard
the elegance of their forms or the beauty and
brilliancy of their colours.  My readers may
perhaps, however, be as much surprised as I was,
to learn that the beautifully coloured parts of
flowers are the least important; and that, as they
only serve as a covering to the stamens and
pistil, which are designed for the production of
seed, they may be, and indeed actually are,
wanting in a great many of what are considered
perfect flowers.  In examining a flower, therefore, it must be remembered that the production of seed is the object, for which all the curi-

ous contrivances we discover are designed. The
germen or ovary (*a* in fig. 1) is protected by a
thick fleshy substance (*b*), called the receptacle
or disk, which serves as a bed or foundation on
which the other parts of the flower rest, and
which is thence frequently called a thalamus or
torus, both words signifying a bed.  The ovary
itself is hollow, and it is sometimes
divided into several cells, each inclosing a number of ovules, which
are afterwards to become seeds ; but
sometimes there is only one cell, and
sometimes only one seed in each cell.
The ovary is juicy and succulent
when young, and very different from

FIG. 1.—STAMEN
AND PISTIL.

what it afterwards becomes when the seeds
are ripe.  Rising from the ovary in most flowers,
is a long and slender stalk called the style
(*c*), which supports a kind of head, called the
stigma (*d*).  The ovary, the style, and the
stigma, constitute what is called the pistil ; but
the style is not so essential as the other parts,
and indeed it is wanting in many flowers.
Sometimes there are many styles, each with
a stigma at its summit, forming the pistil;
and when this is the case, the ovary will have
as many cells as there are stigmas, or each
stigma will have a separate ovary to itself.
There are generally several stamens in a flower,

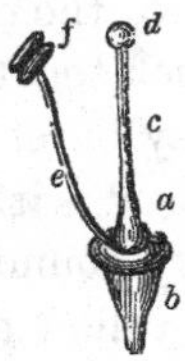

each perfect stamen consisting of three parts,—
the Filament, the Anther, and the Pollen.
The filament (*e*) is, however, often wanting, and
it is only the anther (*f*), and the powder called
the pollen which it contains, that are essential.
The anther, when the flower first expands, appears like a little oblong case with a deep groove
down the centre, or rather like two oblong cases
stuck together.   When these cases become ripe,
they burst and let out the pollen which was inclosed within them.   The pollen is generally
very abundant, and it is often seen in the form
of yellow dust descending from the catkins of
the cedar of Lebanon, or the Scotch fir, or of
orange powder, as on the stamens of the orange
lily, when it sticks to everything it touches.
About the time of the bursting of the anthers,
the stigma becomes covered with a glutinous
moisture, which absorbs the pollen that falls
upon it.   The pollen, when absorbed by the
stigma, is conveyed down the style to the ovary,
where it falls upon and fertilises the ovules or
incipient seeds.   Nothing can be more beautiful
or more ingenious than the mechanism by which
this process is effected.   It is necessary that the
grains of pollen should be separated before they
reach the ovary, and they are so in their pas-
sage down the style in a manner more fine and
delicate than could be done by any exertion of

mere human skill.   We know that we ourselves
are "fearfully and wonderfully made," but how
few of us are aware that every flower we crush
beneath our feet, or gather only to destroy,
displays as much of the Divine care and wisdom
in its construction, as the frame of the mightiest
giant !

I have already mentioned that the most conspicuous part of the flower is merely a covering to protect the seed-producing organs from
injury.   In most flowers there are two of these
coverings, which form together what is called
the perianth; the inner one, when spoken of
separately, being called the corolla, and the
outer one the calyx.   The corolla is generally
of some brilliant colour, and in most cases it is

FIG. 2.—COROLLA OF A
FLOWER.

FIG. 3.—CALYX OF A
FLOWER.

divided into several leaf-like parts called petals,
(see *g* in figs. 2 and 3); and the calyx, which
is commonly green, is divided into similar portions called sepals (see *h*).   Sometimes there is

only one of these coverings, and when this is the
case it is called by modern botanists the calyx,
though it may be coloured like a corolla; and
sometimes the calyx and corolla are of the same
colour, and so mixed as hardly to be distinguished from each other, as in the crocus and
the tulip; in which case the divisions are called
the segments of the perianth.

# CHAPTER I.

THE ORDER RANUNCULACEÆ : ILLUSTRATED BY THE RANUNCULUS, THE BUTTERCUP OR CROWFOOT, THE PEONY, THE ANEMONE, THE HEPATICA, THE CLEMATIS, THE CHRISTMAS ROSE, THE WINTER ACONITE, MONKSHOOD, THE LARKSPUR, AND THE COLUMBINE.

SUCH of my readers who may have formed their first ideas of the natural system from some order, the flowers of which bear a strong resemblance to each other, will be surprised at reading the names of the heterogeneous assemblage of plants at the head of this chapter; for surely no flowers can bear less resemblance to each other than the buttercup and the peony do to the columbine and the larkspur. There are, however, striking points of resemblance which link these flowers together; the principal of which are the number and disposition of the ovaries, or carpels as they are called in this case, which, though they grow close together, and sometimes even adhere to each other, are yet perfectly distinct; in the number and position of the stamens, which grow out of the receptacle from beneath the carpels; and in the leaves and

young stems, when cut or pressed, yielding a thin
yellowish juice, which is extremely acrid, and,
in most cases, poisonous.  The flowers of the
plants belonging to Ranunculaceæ differ widely
in their shapes ; and all the incongruities that
are only sparingly met with in other orders, are
here gathered together.  Some of the flowers
have only a coloured calyx, as in the clematis ;
in others the calyx and corolla are of the same
colour, as in the globe-flower, or so intermingled
as to seem all one, as in the columbine ; and in
others the calyx forms the most ornamental
part of the flower, as in monkshood and the
larkspurs.  In short, modern botanists seem to
have placed this unfortunate order first, as
though to terrify students on the very threshold
of the science, and to prevent them from daring
to advance any farther to penetrate into its
mysteries.

### THE GENUS RANUNCULUS.

THE word Ranunculus will doubtless conjure
up in the minds of my readers those very showy,
double, brilliantly-coloured flowers, which flower
in spring, and are generally grown in beds like
tulips.  These flowers form a species of the
genus, under the name of *Ranunculus asiaticus ;*
and having been introduced from Asia, they
have retained their botanic name from not

having any English one.  The honour of giving
a name to the genus does not, however, rest on
them, but belongs to a common English weed.

Every one who has travelled through England
in the months of June and July, must have
remarked the almost innumerable buttercups
which glitter among the long grass of the meadows at that season ;  and those who observe
closely,  will have noticed that these brilliant
little flowers are never found in poor soil, or in
hilly situations, but in rich valleys where the
grass is rank and luxuriant from abundance of
moisture.  It is this circumstance that has obtained for the buttercup the botanical name of
Ranunculus, the word being derived from Rana,
a frog, a creature that delights in moist places.

The buttercup being the type of the genus
Ranunculus, and the order Ranunculaceæ, a
close examination of its flowers will show the
peculiarities which distinguish both the genus
and the order.  The characteristics of the order,
as far as regards the number and position of the
carpels and stamens, are shown in the section of
the flower in the lower part of *fig.* 4 ;  and those
of the genus are, a green calyx of five sepals, and
a bright coloured corolla of five petals (see *a* in
*fig.* 4) ;  numerous stamens, the anthers of which
are adnate, that is, with the filament growing
up the back (see *b*) ;  and numerous carpels (*c*)

affixed to the upper part of the receptacle, which is drawn up in the shape of a cone to receive them. The flower shown in *fig.* 4, and the detached petal (*e*), given separately to show the little scale at

its base, are of the natural size; but the anther *b* is magnified to show the curious manner in which it is affixed, for its whole length, to the filament. The section of the flower is also magnified to show the elevated receptacle, and the

FIG. 4.—THE FLOWER OF THE COMMON BUTTERCUP.

position of the carpels *c* and the stamens *d* with regard to each other. The line *g* shows the position of the corolla, and *f* that of the calyx, while the short line between the corolla and the stamens indicates the scale, which, from its being supposed to serve as a receptacle for honey, is sometimes called the nectary. The carpels, it will be observed, each consists of a broad part swollen in the centre, which is the ovary, with a curved part or beak at one end, terminating in a sharp point, which is the stigma. Each

ovary contains only one ovule, and when the
seed ripens, the carpel does not open to discharge
it, but drops with the seed.   When the flower
is fully expanded, the green carpels may be seen
in the centre, surrounded by the stamens, as
shown at *h* in *fig.* 5 ; but after the petals drop,

FIG. 5.—FLOWER AND RIPE CARPELS OF THE BUTTERCUP.

the stamens also disappear, and the carpels increase in size, till they assume the appearance
shown at *i*, which shows the kind of head formed
by the carpels on the receptacle after the flower
has faded.

The plant from which my drawings were made
was a common buttercup, *Ranunculus acris*,
which my readers will easily recognise if they
should meet with it, by its erect flower-stem,
deeply cut leaves, and fibrous root.   Another
species (*Ranunculus bulbosus*) is, also, sometimes
called the buttercup; but it is easily distinguished by its bulbous root.   Both these, and
several other species, have deeply cut leaves,

which somewhat resemble the feet of a bird, and
hence the name of crowfoot is often applied to
them.   Others, such as the greater spearwort
(*Ranunculus lingua*), have long tongue-shaped
leaves.   In all, the footstalks of the leaves are
somewhat folded round the stem at their base.
Such of my readers as reside in the country
will find it very amusing to gather all the kinds
of crowfoot, buttercup, goldilocks, and spearwort,
they can find in the fields and lanes ;  and after
having compared the flowers with the description
I have given, to try to find out the specific
names, by comparing the other particulars with
the   descriptions   in   Hooker's   or   Lindley's
British Botany, or with the plates and descriptions in the new edition of Sowerby's English
Botany.   In a short time they will not want
these aids, but will be able to name the plants
at once, and to tell in what they differ from each
other by memory.   I shall never forget the
pleasure I once had in finding out the name of
a plant myself.   I happened to be waiting for
Mr. Loudon, (who had gone to examine some
new pines and firs,) in the pleasure-grounds of
a villa, just opposite a small pond, which was
covered by some white flowers that I did not
know.   The flowers were small, but very beautiful, and as they shone with almost a metallic
lustre in the sun, they looked like a  silvery

mantle thrown over the water. I was curious
to know what they were, and having got one
with some difficulty, and by the help of my
parasol, I began to examine it botanically. The
leaves at first told me nothing as to the genus,
for the upper ones were nearly round, and only
slightly cut into three lobes, while the lower ones
were almost as much divided as fennel; but on
examining them closely, I found their stalks
sheathed the stem at the base. This gave me
the first idea of the plant being a Ranunculus,
for I remembered the leaves of that genus were
stem-clasping. I then looked at the plant again,
and wondered at my own stupidity in not having
before observed its resemblance to the genus.
There was the cup-shaped flower-of five petals,
the green calyx of five sepals, the numerous
stamens and carpels, the elevated receptacle,
and even the fine texture and glossy surface of
the petals. Nothing was different but the colour;
and yet it was the want of the bright golden
yellow of the common buttercup, that prevented
me from even thinking of that genus, when pon-
dering on the name of my water-plant. I should
add, that I would not ask any help from Mr.
Loudon, but identified my plant myself on my
return home; when, by comparing it with the
description in Hooker's British Flora, which
happened to be the first botanical work I had

at hand, I found it was *Ranunculus aquatilis*, the water crowfoot.

In a similar manner my readers may amuse themselves, by identifying the plants they meet with, and they will be surprised to find how easy the task will soon become. I must warn them, however, that they will not find double flowers quite so easy to recognise as single ones. In double flowers the stamens and carpels are entirely or partially changed into petals; as may be seen in the florists' varieties of Ranunculus, in the yellow bachelor's buttons, which is a variety of the common buttercup, and in the Fair Maid of France, which is a variety of *Ranunculus platanifolius*, a species found wild on the mountains of Germany.

### THE GENUS FICARIA.

Pansies, lilies, kingcups, daisies,
Let them live upon their praises ;
Long as there's a sun that sets,
   Primroses will have their glory ;
Long as there are violets,
   They will have a place in story.
There's a flower that shall be mine,
'Tis the little Celandine.

Ill befall the yellow flowers,
Children of the flaring hours,
Buttercups that will be seen,
   Whether we will see or no ;
Others, too, of lofty mien,
   They have done as worldlings do,
Stolen praise that should be thine,
Little humble Celandine.

In these verses, and several others in the same strain, Wordsworth sings the praises of the pretty little British plant called pilewort, or the lesser celandine. This plant botanists formerly included in the genus Ranunculus, but De Candolle, finding that instead of having five sepals and five petals like all the kinds of Ranunculus, it has three sepals, and nine petals, which are narrow and pointed, instead of being broad and somewhat rounded, made it into a new genus under the name of Ficaria ranunculoides—its old name having been Ranunculus Ficaria. Its flowers are of a bright yellow, like those of the buttercup, and of the same delicate texture and glossy surface; but they are distinguished, not only as I have just observed, by having nine narrow pointed petals, and only three sepals, but by the leaves, which are roundish and shining, and not stem-clasping. These peculiarities are so striking, that I knew the Ficaria the first time I saw it in a growing state, merely from having read a description of it. Even when not in flower it may be known, by its roundish smooth leaves, and by the petioles or footstalks of its leaves being the same throughout; whereas those of all the kinds of Ranunculus are dilated at the base, to enable them to enfold the stem.

c

## THE GENUS PÆONIA.

The flowers of the Peony bear considerable resemblance to those of the buttercup, but every part is on an enlarged scale ; and there are some important differences—one of which is, that the Peony retains its calyx till the seeds are ripe, while in all the kinds of Ranunculus the calyx

Fig. 6.—Flower of the male Peony, with detached carpel and stamen.

drops with the corolla. The carpels of the Peony are also many-seeded, while those of the Ranunculus contain only one seed in each. In the male Peony (*P. corallina*) there are five petals and five sepals, (see *a* in *fig.* 6,) with numerous stamens, forming a ring round four large woolly carpels in the centre of the flower. The stamens (*c*) are adnate, like those of the Ranunculus ; and the carpels (*b*) are each terminated by a thick, fleshy, hooked stigma. These carpels open naturally on the side when ripe, to discharge their seeds. The herbaceous Peonies with double flowers, now so common in our gardens, have generally only two carpels, each containing about twenty seeds, arranged in two rows ; and the Chinese tree Peony (*P. Moutan*) has from five to ten carpels, with only a few seeds in each. This last species is distinguished by the receptacle being drawn out into a thin membrane-like substance, which rises between the carpels like the remains of withered leaves, and partially covers them.

### THE GENUS ANEMONE.

I HAVE already mentioned (p. 10) that some of the genera included in the order Ranunculaceæ have only a coloured calyx and no corolla; and the Anemone is an example of this peculiarity of construction. The pasque-flower (*Anemone pulsatilla*) is divided into six dark purple sepals, which are covered on the outside with long silky hairs. The leaves are so much cut as almost to resemble those of parsley ; and at a short distance below the flowers there are three small floral leaves, or bracts, which grow round the stem, and form what is called an involucre. The carpels are small, oblong bodies, pressed close together, and each is furnished with a long, feathery point, called an awn. The

carpels, though lying so close together, are
perfectly distinct, and part readily at the
slightest touch; and each contains only one seed.

It will be seen from this hasty sketch, that
the principal point of resemblance between the
genera Anemone and Ranunculus, in a botanical
point of view, lies in the carpels, which are close
together, and are yet so distinct as to part at
the slightest touch. There is, however, a general
resemblance in some of the flowers, from their
five sepals, and numerous stamens, that renders
it difficult for a beginner to distinguish an Anemone from a Ranunculus. In many of the British
species, also, the carpels are not awned, but
slightly curved, very like those of a buttercup.
I remember being once very much puzzled with
a beautiful little bright yellow flower, that I
found in a wood. At first I thought it was a
Ranunculus, but the petals were pointed and
not roundish; and it could not be a Ficaria,
because it had only five petals. At last I looked
to see what kind of calyx it had, and found
none, that is, no green calyx; and then, ob-
serving the involucre of three leaves growing in
a whorl round the stem, at some distance below
the flower, I knew it was an Anemone; and on
comparing it with the plates in Sowerby's
English Botany, on my return home, I ascertained that it was *Anemone ranunculoides*.

My readers will therefore observe that Anemones may be always known by their involucre,
and by their having only one covering (a showy,
coloured calyx) to the flower.  The number of
sepals in this calyx varies in the different species.
The pasque-flower has six; the white wood
Anemone generally five; and the Blue Mountain
Anemone from twelve to twenty.  The involucre also sometimes grows a long way from
the flower, as in this last-mentioned species;
and sometimes so close to it, as in the Garland,
or Poppy Anemone (*A. coronaria*), as to look
almost like a green calyx to the flower.  The
awns, or feathery tails, are also not found
attached to the carpels of all the species; and
this distinction is considered so important, that
some botanists make those plants which have
awned carpels into a separate genus, which they
call Pulsatilla, and of which the pasque-flower
is considered the type.  This genus, however,
has not, I believe, been generally adopted.

I have now only a few words to say on florists'
Anemones, the tuberous roots of which most of
my readers must have seen in the seed-shops.
Most of these are varieties of the Garland
Anemone, already mentioned as having its involucre close to the flower.  The sepals of this
species are roundish, six in number, and when
the flower is in a single state, there are a great

number of stamens, bearing dark purple anthers
in the centre of the flower. When the flower
becomes double, the sepals, which retain their
form and number, only becoming somewhat
more spread out and flattened, are called by
florists the guard-leaves; and the stamens in
the centre are metamorphosed into petals,
which generally retain their dark purple colour,
or at any rate are much darker than the sepals.
The other florists' Anemones spring from *A.
stellata*, or *hortensis*, and they are distinguished
by having pointed sepals, and a white spot at
the base of each, so as to form a white circle
inside the cup of the flower. The involucre is
a long way from the calyx, and when the flowers
become double, the sepals can scarcely be distinguished from the metamorphosed stamens.

The hepatica or liverwort, the varieties of
which look so pretty in our gardens in spring,
was formerly considered to be a species of Anemone, and indeed the genus Hepatica appears
to rest on very slight grounds. It has, however,
been adopted by most modern botanists, and
the *Anemone Hepatica* of Linnæus is now generally called *Hepatica triloba*. The normal form
of the species is the single blue; and the double
blue, the single and double pink, and the single
and double white, are all only varieties of this.
The hepatica agrees in all points with the

Anemone, except in the involucre, which is so very like a green calyx, from the manner in which it enfolds the flower in the bud, as scarcely to be distinguished. I could not, indeed, be persuaded that this calyx-like covering was an involucre, till I turned back the apparent sepals, and found that their glossy surface was within: I also found that there was a very small portion of the stem between them and the flower, a circumstance which always distinguishes an involucre from a calyx, the latter forming part of the flower, and being always in some manner attached to the receptacle.

## THE GENUS CLEMATIS.

This genus resembles the Anemone in having only one covering, an ornamental calyx, to its seed-producing organs. It has not, however, any distinct involucre; though in one species, *C. calycina*, there are two bracts, or floral leaves, which bear some resemblance to one. The flowers of the different species vary considerably in form, colour, and the number of the sepals; *C. calycina* and *C. viticella* having four, *C. florida* six, *C. vitalba* five, &c. All the species agree, however, in the seeds, which are produced singly, each in a separate awned carpel, which does not open, but drops with the seed, and is sown with it. These carpels, which are common

to the genera Ranunculus, Anemone, Adonis,
and many other kinds of Ranunculaceæ, are
called caryopsides, and seeds thus enclosed are
always much longer in coming up than any
others. In some species of Clematis the awns
of the carpels are smooth; but in others they

FIG. 7.—CARPELS OF THE CLEMATIS VITALBA.

are bearded or feathered, as in those of the
traveller's joy (*C. vitalba*), shown in *fig.* 7. The
leaves of the Clematis vary considerably in form
and arrangement; but the stems of the climbing
species are furnished with tendrils, or slender
twining leafless stems, which some botanists suppose to be metamorphosed leaves.

The plants composing the genus Atragenè
have been separated from Clematis; because
they are said to have petals, which the genus
Clematis has not. It must not, however, be
supposed that the petals of the Atragenè bear

any resemblance to what is generally understood
by that word.    On the contrary, the showy part
of the Atragenè is still only a coloured calyx ;
while the petals are oblong, leaf-like bodies in
the centre of the flower, which look like dilated
stamens.    In other respects the two genera are
scarcely to be distinguished from each other.

### THE GENUS HELLEBORUS, &c.

THE Christmas rose (*Helleborus niger*) bears
considerable resemblance in the construction of
its flowers to the Atragenè, for it has a showy
calyx, and narrow oblong petals, encircling the
stamens in the centre of the flower.    The calyx
of the Christmas rose is white, delicately tinged
with pink, and the petals are green.    The
carpels are erect and long, swelling out at the
base, and each ends in a curved style with a
pointed stigma.    The Christmas rose takes its
specific name of niger (black) from the root,
which is covered with a thick black skin.    The
common Hellebore takes its name of *H. viridis*,
from its flowers, which are green.    The carpels
of this plant frequently grow slightly together,
and their styles curve inwardly.

The British species of Hellebore have no involucre, and the Christmas rose has only two
bracts or floral leaves, which form a calyx-like
covering to the bud ; but the little yellow

garden plant, called the Winter Aconite, which
was included by Linnæus in the genus Helleborus, has a decided involucre, on which the
little yellow, cup-shaped flower reposes, like a
fairy bowl upon a leafy plate. The conspicuous
part of this flower, like the others, is the calyx,
which encloses a number of short tubular petals.
This little plant is now separated from Helleborus, and formed into a distinct genus, under
the name of *Eranthus hyemalis*, from its carpels
being each furnished with a very short footstalk, by which they are attached to the receptacle, instead of growing upon it as in the other
genera. The root is tuberous, or rather it
forms a kind of underground stem, sending up
tufts of leaves and flowers from the different
buds. Thus we often see several tufts of the
Winter Aconite growing so far from each other as
to appear distinct ; but which, in fact, all spring
from the same root. The Globe-flower (*Trollius
europæus*), which has a golden yellow, globe-
shaped calyx, enclosing a number of small oblong
petals, is nearly allied to the Winter Aconite ;
and the Fennel-flower, or Devil in a Bush (*Nigella
damascena*), agrees with the common Hellebore
in the adhesion of its carpels.

### THE GENUS ACONITUM.

WE are so accustomed to see in our gardens
the tall showy perennial called monkshood or

FIG. 8.—FLOWER AND SEED-VESSELS OF THE MONKSHOOD.

wolfsbane (*Aconitum Napellus*), that few persons
think of examining the flowers in detail.  They
well deserve, however, to be examined, as they
are very curious in their construction.   The
showy part of the flower is an ornamental calyx
of six sepals, but the upper two of these are

larger than the others, and adhere together so
as to form a singular sort of covering, like a
monk's cowl or hood. (See *a* in *fig*. 8.)  The
stamens are numerous, and they encircle three
or five oval carpels, with thread-like styles, and
pointed stigmas, as shown at *b ;* which when
ripe burst open at the top (*c*) to discharge the
seed, without separating.  Carpels of this kind
are called follicles.  Under the hood, and entirely concealed by it, are the petals (see *fig*. 9),

which form what may certainly be
considered the most remarkable
part of the flower, as they are so
curiously folded up that they look
more like gigantic stamens than
petals.  The older botanists described these petals as nectaries,
with crested claws.  The leaves
are divided into from three to five
principal segments, which are again deeply cut
into several others.  The stem of the common
Monkshood is thickened at the base, or collar,
where it joins the root, so as to give it somewhat the appearance of celery; and hence
ignorant persons have been poisoned by eating
it.  This knotted appearance of the stem is not
common to all the species, and it gives rise to
the specific name of Napellus, which signifies a
little turnip.

FIG. 9.—PETALS OF
THE MONKSHOOD.

## THE GENUS DELPHINIUM.

THE plants belonging to the genus Delphinium, that is to say, the Larkspurs, have their flowers constructed in nearly as curious a manner as those of the different kinds of Monkshood ; but they differ in the sepals and petals both forming

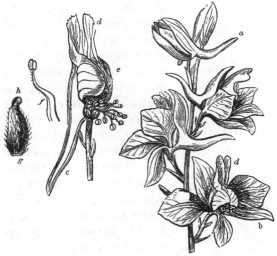

FIG. 10.—THE FLOWERS OF THE BRANCHING LARKSPUR.

conspicuous parts of the flower, though they are generally quite distinct both in form and colour, and may be easily traced through all the different forms they assume in the various species. They are, however, perhaps most easily distinguished in the branching or autumnal Larkspur (*Delphinium consolida*).  In the flower of this plant the spur (*a* in *fig.* 10) is the upper sepal of the calyx, and it serves as a cover to part of the petals.   There are four other portions of the calyx (*b*), which assume the appearance of ordinary sepals.  The petals are four in number;

and they are united at the lower part, and drawn out into a sort of tail, as shown at *c ;* while the upper part of two of them stands up like asses' ears (*d*) in the centre of the flower ; and the others are curiously folded, so as to form a hood over the stamens and carpels, as shown at *e.*  The anthers of the stamens resemble those of the Ranunculus; but the filaments are bent, as shown at *f.*  The carpels (*g*) are upright, hairy, and terminate in a blunt, fleshy stigma (*h*). When ripe, they open in the same manner as those of the Monkshood.  The branching Larkspur has a fusiform or tap root, as shown in *fig.* 11, in which *a* is the

Fig. 11.—Tap root of the Branching Larkspur.

collar, or as the Italians call it *la noda vitale;* and *b* the fibrous roots, through the points of which the plant takes up its food.

The flowers of the other kinds of Larkspur resemble this one in their general appearance, though they differ in the minor details. Those of the Rocket Larkspur (*D. Ajacis*) lose their spurs when they become double; and those of the Bee Larkspurs have their petals nearly black, and instead of standing up like ears, they are so curiously folded as to resemble a bee nestling in the centre of the flower.

### THE GENUS AQUILEGIA.

THE common Columbine (*Aquilegia vulgaris*) differs from all the flowers I have yet described in having the sepals and petals not only of the same colour, but so intermingled as to be scarcely distinguishable from each other. The flower (given on a reduced scale at *a* in *fig.* 12) is composed of five horn-shaped petals, which are curved at the upper end, and form a kind of coronet round the stem; and five oval sepals, which are placed alternately with them; all, generally speaking, being of the same colour. The horn-shaped petal, or nectary as it was called by Linnæus, is attached to the receptacle at the thickened rim (*b*), while the sepal is attached at the point (*c*); *d* shows the dis-

position of the stamens; *e* a separate stamen,
with its adnate anther; *f* the inner row of

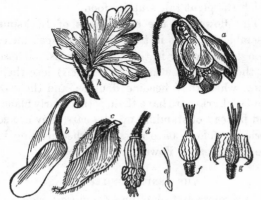

FIG. 12.—FLOWER AND LEAF OF THE COLUMBINE.

stamens, which are produced without anthers,
and with their filaments growing together, so
as to form a thin membranaceous case for the
carpels, which are shown exposed at *g*. The
carpels, when ripe, become follicles. The leaf
of the Columbine is bi-ternate; that is, it is cut
into three large divisions, each of which is cut
into three smaller ones; so that it is twice-
ternate. The petiole or foot-stalk of the leaf
sheaths the stem, as shown at *h*, where the leaf
is represented on a reduced scale to suit the
flower.

I would advise such of my readers as are
anxious to turn the preceding pages to account,
to procure as many of the plants I have described
as possible, and to compare them with each
other, and with any other plants belonging to
the order Ranunculaceæ that they can obtain.
Those who have access to a botanic garden
will have no difficulty in finding the names of
the genera included in the order; and those
who have not this advantage, must consult
Don's edition of Sweet's Hortus Britannicus, or
any other catalogue in which the plants are
arranged according to the Natural System.
When a number of specimens have been collected, the student will be surprised to see how
many points of resemblance exist between them.
The stems of all, when cut, will yield a watery
juice; which is always acrid, though some
of the plants are more poisonous than others.
The stamens will be found to be always numerous, and always attached to the receptacle
below the carpels; and the anthers are generally
adnate, that is attached to the filaments from one
end to the other (see p. 12). The carpels are in
most cases numerous, and either distinct, or
adhering in such a manner as to show plainly
the line of junction between them; they are
also always one-celled, whether one or many-
seeded, and generally either caryopsides (see

D

p. 24), or follicles (see p. 28). The leaves are
generally divided into three or five lobes, each
of which is cut into several smaller divisions;
and the petioles or leaf-stalks are very frequently
dilated at the base, and sheathing the stem.
In most cases, the flowers are of brilliant colours,
several of them being cup-shaped, and many
with the calyx more ornamental than the corolla.
The seeds will generally keep good for several
years; and several of them, particularly those
of the kind called caryopsides, when sown, are
often a long time before they come up.

# CHAPTER II.

THE ORDER LEGUMINOSÆ: ILLUSTRATED BY THE SWEET-PEA, THE RED CLOVER, ACACIA ARMATA, THE SENSITIVE PLANT, THE BARBADOES FLOWER-FENCE, THE CAROB-TREE, THE TAMARIND, THE SENNA, THE GLEDITSCHIA, THE LOGWOOD, THE JUDAS-TREE, AND THE KENTUCKY COFFEE-TREE.

THIS order is a very numerous one, containing above three hundred genera, and including several highly important plants, both for food and commerce. As examples of the utility of the Leguminosæ for food, I need only mention the pea and bean, and all their numerous allies; and as examples of their importance in medicine and the arts, I may enumerate senna, liquorice, the tamarind, gum-arabic, and logwood. Among the ornamental plants belonging to this order are, the Laburnum, the Furze or Gorse, the Robinia or False Acacia, the true Acacias, the Sensitive Plant, and the Barbadoes Flower-fence. It will be seen by this enumeration, that the flowers of the Leguminosæ differ from each other nearly as much as those of Ranunculaceæ; but when in seed, they are all easily recognised by their seed-vessels, which are always legumes, that is, bearing more or less resemblance to the pod of the common pea. To aid the memory

in retaining the great number of genera included
in this order, various methods have been devised
of re-dividing it ; and of these I shall adopt the
newest, which is also the simplest, by which they
are arranged in three tribes, according to their
flowers.

----

### TRIBE I.—PAPILIONACEOUS FLOWERS.

THE flowers of this tribe are called Papilionaceous ; because
Papilio is the sci-
entific name of a
genus of butter-
flies, which they
were supposed to
resemble.      The
type of this tribe
may be considered
the flower of the
sweet-pea  (*La-
thyrus odoratus*),
which has a small
green calyx, cut
into   five   deep
notches, but not
divided into regu-
lar sepals. (See *a*
and *b* in *fig.* 13.)

FIG. 13—FLOWER, POD, AND TENDRIL, OF
THE SWEET-PEA.

The corolla is in five petals, the largest of

which (c) stands erect, and is called the vex-
illum or standard ; below this are two smaller
petals (d), which are called the algæ or wings ;
and below these are two petals, joined together
so as to form a kind of boat (e), which are called
the carina or keel, and which serve as a cradle
for the stamens and pistil.    There are ten
stamens, nine of which have the lower half of
their filaments growing together, so as to form
a fleshy substance at the base, as shown in *fig.*
14 at *f,* and the other (g) is free.
The ovary is oblong, terminating
in a filiform style, with a pointed
stigma, as shown at *g* in *fig.* 13 ;
and it is one-celled and many-
seeded ; the seeds being what we
call the peas.    When the petals
fall, the pod still retains the calyx
(b), and the style (g) ; and these remain on till
the seeds are ripe, when the pod divides naturally into two parts, or valves as they are
called, which curl back so as to discharge the
seeds.    If the pod be examined before it bursts,
it will be found that the valves are composed
of a fleshy substance, lined with a strong membrane or skin, and that they are united by two
seams, called the dorsal and ventral sutures.
Along the ventral suture (h) there runs a kind
of nerve, called the placenta, to which the peas

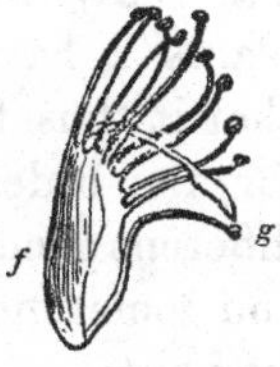

FIG. 14.—STAMENS
OF THE SWEET-PEA.

are attached, each pea being furnished with a
little separate stalk, called a funicle.  A cook
would be surprised, even in these enlightened
times, to be told to take a legume of *Pisum
sativum*, and after separating the two valves at
the dorsal suture, to detach the funicles of the
seeds from the placenta ; yet these scientific
terms would merely describe the operation of
shelling the peas.  It will be seen by this description that the pod of the pea differs very
materially from the seed-vessels of all the other
plants I have had occasion to describe ; and
that it thus forms a very distinctive character
for the order.  The other parts vary in the
different genera: the calyx is sometimes tubular,
and sometimes inflated ; sometimes it has only
four notches, or teeth as they are called, instead
of five, and sometimes it has five distinct sepals
divided to the base.  The parts of the corolla
vary also in proportion to each other, the keel
in some of the Australian plants is as long as
the standard ; as, for example, in *Kennedia
Maryattæ ;* and in others the wings are so small
as to be scarcely visible.  The stamens of many
of the species are also free, that is, divided to
the base ; while in others they resemble those
of the sweet-pea, in having nine joined together
and one free ; and in others the whole are
joined together at the base.  The pods also

vary very much in size and form; being sometimes nearly round, and only one or two-seeded;
and in others long, and containing many seeds,
as in the common bean or pea. The seeds
themselves are so different that the tribe has
been divided, on account of them, into two
sections : the one consisting of those plants
which, like the common bean, have the seed
dividing into two fleshy seed-leaves or cotyledons,
when it begins to germinate ; and the other,
the seed-leaves of which are thin. The seeds of
the papilionaceous plants which have thin cotyledons are not eatable ; but those with fleshy
cotyledons may be safely used as food. The
fleshy cotyledons do not always rise above the
ground; but they do so decidedly in the bean
and the lupine ; and if either of these seeds be
laid in moist soil with the hilum or scar downwards, the seed, as soon as it begins to germinate, will divide into two parts (that is, into
two cotyledons), which will rise above the
ground, and become green like leaves ; though,
from still retaining their roundish form, they
are easily distinguished from the true leaves,
which rise in the centre. Though my readers
will have no difficulty in recognising most of the
Leguminosæ which have papilionaceous flowers,
there are some genera, respecting which they
may be interested to learn a few particulars.

Thus, the Chorozema is one of the kinds with thin cotyledons, and consequently its seeds are not eatable. The legumes of this genus are roundish, and swelled out, so as to bear but little outward resemblance to a pod. Sophora, Edwardsia, Virgilia, Podolobium, Callistachys, Brachysema, Burtonia, Dillwynia, Eutaxia, Pultenæa, Daviesia, and Mirbelia, have all thin cotyledons, and their ten stamens all separate from each other; but in Hovea, Platylobium, and Bossiæa, though the cotyledons are thin, the stamens all grow together at the base. I mention these common greenhouse shrubs, that my readers may have an opportunity of examining their botanical construction, and thus verifying their names. The common furze (*Ulex europæus*), the Spanish broom (*Spartium junceum*), the Petty whin (*Genista Anglica*), the Laburnum (*Cytisus Laburnum*), and the common broom, all belong to this division, and consequently their seeds are not eatable; those of the Laburnum are indeed poisonous. The distinctions between Spartium, Genista, and Cytisus, are very slight, lying chiefly in the calyx; and as a proof of this the common broom, which is now called *Cytisus scoparius*, was formerly supposed to be a Spartium, and afterwards a Genista.

The common red clover (*Trifolium pratense*)

has its flowers in such dense heads that it is
difficult at first sight to discover that they are
Papilionaceous.  On examination, however, it
will be found that each separate flower has its
standard, wings, and keel, though the wings
are so large as to hide the keel, and nearly to
obscure the standard.  The calyx is tubular at
the base, but divided above into five long, awl-
shaped teeth, that stand widely apart from each
other.  The legume has only one or two seeds,
and it is so small as generally to be hidden by
the calyx.

## TRIBE II.—MIMOSÆ.

THE second division of Leguminosæ comprises those plants which have heads of flowers
either in spikes or
balls, like those
shown in *fig.* 15.
This figure repre-
sents two heads of
flowers of *Acacia ar-
mata*, a well known
greenhouse shrub,
of their natural
size ; and *fig.* 16
shows a head of
similar flowers mag-
nified.  In the lat-

FIG. 15.—FLOWERS AND SPRIG OF
ACACIA ARMATA.

ter, *a* shows the calyx, which is five-toothed, and *b* the petals, which are five in number and quite regular in shape; *c* are the stamens, which vary from ten to two hundred in each flower, and which are raised so high above the petals as to give a light and tuft-like appearance to the whole flower.

FIG. 16.—FLOWER OF ACACIA MAGNIFIED. The legumes are very large in proportion to the flower; and consequently, by a wise provision of nature, only a very few of the flowers produce seed. The valves of the legumes are not fleshy like those of the pea, but dry and hard, and when they open they do not curl back.

The flowers in the different kinds of Acacia, differ in the corolla, which has sometimes only four petals, which are occasionally united at the base, and in the calyx, which is sometimes only four-cleft. The flowers also in many species are in spikes instead of balls.

The rest of the plant of *Acacia armata* is very curious; what appear to be the leaves (see *d* in *fig.* 15) are, in fact, only the petioles of the leaves dilated into what are called phyllodia; the true leaves, which were of the kind called bi-pinnate, having fallen off, or never unfolded. The true leaves, however, often appear on seedling plants; and thus, when seeds are sown

of several kinds of Acacia, it is sometimes difficult to recognise them till they have attained

FIG. 17.—THE BI-PINNATE LEAF OF AN ACACIA.

a considerable age.  The stipules of the leaves,
(which are to ordinary leaves what bracts are
to flowers,) are in *Acacia armàta* converted into
spines, as shown at *e*.  In some kinds of Acacia
the true leaves, with the petioles in their natural
state, (see *fig.* 17,) are retained in the adult
plants, as in *Acacia dealbàta ;* and in others, the
bi-pinnate leaves are occasionally found attached
to the phyllodia, as in *A. melanoxylon.*  The
bi-pinnate leaves are composed of from six to
twenty pairs of pinnæ, or compound leaflets (see
*f* in *fig.* 17), each of which consists of from eight
to forty pairs of small leaflets (*g*).  The Gum
Arabic tree, *Acacia vera*, has leaves with only
two pairs of pinnæ, but each has eight or ten
pairs of small leaflets.  The branches and spines
are red, and the heads of flowers are yellow.

There are above three hundred known species of Acacia.

The genus Mimosa differs from Acacia in the corolla being funnel-shaped, and four or five cleft. There are seldom above fifteen stamens, which are generally on longer filaments than those of the Acacia; and the legume is compressed and jointed or articulated between the seeds, so that the part which contains one seed may be broken off, without tearing the rest. The Sensitive-plant (*Mimosa pudica*) is a familiar example of this genus.

The cotyledons of the plants belonging to this tribe are generally leafy; and the seeds are not eatable. The plants themselves are easily recognised by their ball or tassel-shaped heads or spikes of flowers; by the small cup-shape and inconspicuous corolla of each; by the grea number and length of the stamens; and by their bi-pinnate leaves, or phyllodia supplying the place of leaves—though the phyllodia are sometimes found in Australian plants with papilionaceous flowers, as, for example, in *Bossiæa ensata*.

--------

### TRIBE III.—CÆSALPINEÆ.

THE flowers of the plants contained in this tribe have generally five regular, widely spreading petals, which are never joined together;

and stamens of unequal length, which with few
exceptions are also perfectly free.  The petals
are generally of the same size and shape; though
sometimes, as in the Barbadoes Flower-fence
(*Poinciana*, or *Cæsalpinia pulcherrima*), four are
of the same shape, and one deformed (see *fig.* 18).

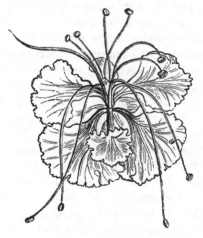

FIG. 18.—FLOWER OF THE BARBADOES FLOWER-FENCE.

The filaments of some of the stamens are very
long and curving over, but the others are much
shorter and erect; the style is long and slender,
ending in a pointed stigma.  The legume is
flat, and it looks almost many-celled, from the
seeds being divided from each other by a kind
of spongy substance, frequently found in the pods

of plants belonging to this division.   The leaves
are bi-pinnate, and the stem is spiny.

The Carob-tree, or St. John's bread (*Ceratonia
siliqua*), agrees with the Barbadoes Flower-fence
in the pulpy matter dividing the seeds, though
it differs widely in its flowers, which are without
petals, and do not possess any beauty.   The
pulp of the pods of the Carob tree is eatable;
but that of Poinciana is said to be injurious.
The pod of the Tamarind (*Tamarindus indica*)
differs from the preceding species in having the
pulpy matter of its pods contained between the
outer and inner skin of each valve, like the
fleshy substance in the pod of the pea, instead
of serving as a bed for the seeds.   The flowers
of the tamarind have five equal petals of a
brownish yellow, three of them being streaked
with pink; and the anthers are nearly rose-
colour.   The stamens and the style both curve
upwards.   It is the pods prepared with sugar
that form what we call Tamarinds.   In *Cassia
lanceolata*, the leaves of which furnish senna, the
flowers have a bright yellow corolla of five concave petals, three of which are somewhat larger
than the others.   The stamens are also unequal
in length; and the style curves upwards.   The
legume is kidney-shaped, and the cells are
divided from each other by thin membraneous
partitions.   The Gleditschia or Honey Locusts,

now so frequently planted in our shrubberies on
account of the lightness and elegance of their
foliage, belong to this division, and some of
them, particularly the Chinese Thorny Acacia
(*Gleditschia horrida*), are remarkable for their
thorns proceeding from the trunk and large
branches, as well as from the axils of the leaves.
The Logwood (*Hæmatoxylon Campechianum*),
has inconspicuous yellow flowers, the petals
being very little longer than the calyx ; and the
legume has seldom more than two seeds. Though
it is considered a tree, the stem is seldom thicker
than the arm of a man, and it is generally
crooked ; chips of the wood are used for dyeing
purple. The Judas-tree (*Cercis siliquastrum*)
is another species belonging to this division, as,
though the flowers appear of the papilionaceous
kind, they are, in fact, composed of five petals,
nearly equal in size, but having the wings the
largest. There are ten stamens, free, and of
unequal length. The legume is oblong and
many-seeded ; and it opens only on the dorsal
suture, the other side to which the seeds are
attached being slightly winged. The flowers
are each on a separate flower-stalk or pedicel,
but they rise from the trunk and branches in
tufts or fascicles. The leaves are simple and
cordate ; and they do not appear till the flowers
have faded.

The Kentucky Coffee-tree (*Gymnocladus canadensis*) is the last plant belonging to this division
that I shall attempt to describe. This tree is
called in Canada, Chicot, or the stump-tree, from
its having no visible buds, and thus appearing
like a dead stump in winter. The flowers of
this plant are white, and they are produced in
racemes, but they bear no resemblance to the
pea flowers, having rather a star-like appearance, like those of the Jasmine (see *fig.* 19).

The calyx (*a*) is tubu-
lar; and the upper
part or limb is divid-
ed into five parts (*b*),
which alternate with
the petals of the
corolla (*c*). There
are ten stamens, but
they are completely
enclosed in the tube of the calyx. The pod is

FIG. 19.—FLOWERS OF THE KENTUCKY
COFFEE-TREE.

very large, the valves becoming hard and bony
when dry; and the seeds are like large beans,
the pod being deeply indented between the seeds.
The leaves are bi-pinnate, with from four to
seven pairs of pinnæ; the lower having only
one small leaflet, but the rest bearing from six
to eight pairs of leaflets each. This tree must
not be confounded with the true Coffee-tree,
which belongs to Rubiaceæ, and from which it is

perfectly distinct in every respect; and it only
takes its American name from its beans having
been used as a substitute for coffee.   The outer
bark of this tree, when it becomes old, splits off
in narrow strips and rolls up; and its timber,
like that of the Robinia or False Acacia, having
very little sap wood, is thus very strong in
quite young trees, though it is of little value
when the tree is full-grown.

The species contained in the first and second
divisions of this order will be easily recognised
by botanical students; and though those of the
third division are much more difficult to find out,
still there is a kind of family likeness, particularly in the leaves, which will enable the eye,
with a little practice, to recognise them.   The
student should visit the hothouses of botanic
gardens and nurseries, and should there endeavour to pick out plants belonging to this order.

## CHAPTER III.

THE ORDER ROSACEÆ, ILLUSTRATED BY DIFFERENT KINDS OF ROSES ; THE POTENTILLA ; THE STRAWBERRY ; THE RASPBERRY ; SPIRÆA ; KERRIA OR CORCHORUS JAPONICA; THE ALMOND ; THE PEACH AND NECTARINE ; THE APRICOT ; THE PLUM ; THE CHERRY ; THE APPLE ; THE PEAR ; THE MOUNTAIN ASH ; THE WHITE BEAM TREE ; QUINCE ; PYRUS OR CYDONIA JAPO-NICA ; THE HAWTHORN ; THE INDIAN HAWTHORN ; THE MEDLAR ; PHOTINIA ; ERIOBOTRYA ; COTONEASTER ; AMELAN-CHIER ; BURNET ; AND ALCHEMILLA OR LADIES'-MANTLE.

ALL the numerous plants which compose this large order agree more or less with the rose in the construction of their flowers, though they differ widely in the appearance of their fruit. They all agree in having the receptacle dilated, so as to form a lining to the lower part of the calyx, and in the upper part of this lining the stamens and petals are inserted above the ovary ; and the anthers are innate, that is, the filament is inserted only in the lower part. The leaves also have generally large and conspicuous stipules ; and they are frequently compound, that is, composed of several pairs of leaflets, placed exactly opposite to each other; though the leaves themselves are never opposite to each other, but are placed alternately on the main stem. These

characters are common to the order; but the
plants included in it differ from each other
so much in other respects, that it has been
found necessary to redivide Rosaceæ into tribes,
of which the following six contain plants common in British gardens.

### TRIBE I.—ROSEÆ.

THE flowers of the wild Rose have the lower
part of the calyx tubular and fleshy (from being
lined with the dilated receptacle) and the upper
part divided into five leafy sepals, which enfold

FIG. 20.—ROSA FOSTERI.

the bud, and remain on after the expansion of
the corolla. In *Rosa Fosteri*, (see *fig.* 20,) and

E 2

its near ally the Dog rose(*R. canina*), the sepals
(*a*) do not extend far beyond the petals of the
bud; but in some species, as in *Rosa cinnamonea*
and its allies, the sepals are so large and long,
that they assume the character of little leaves,
The corolla is cup-shaped, and it is composed of
five equal petals, each of which is more or less
indented in the margin, as shown at *b*.   In
the centre of the flower the receptacle forms a
kind of disk which completely fills the opening or
throat of the calyx; in most species covering the
carpels and their styles and only leaving the stigmas
free, though in the Ayrshire rose (*R. arvensis*),
and its allies, the styles are united, so as to
form a column, which projects considerably above
the disk (see *fig.* 21).

The pitcher - shaped
part of the calyx when
the corolla falls be-
comes the hip ( *fig.*
20 *c*), and serves as
a covering or false
pericarp to the nu-
merous bony carpels

FIG. 21.—OVARY OF THE AYRSHIRE
ROSE WITH A DETACHED SEED.

or nuts which contain  the seed.  These nuts
are each enveloped in a hairy cover (see *fig.*
20 *d*, and *fig.* 21 *a*,) and each contains only
one seed which it does not open naturally to
discharge : hence, the seeds of roses when sown

are a long time before they come up.　*Fig.* 22
is the ripe fruit of *Rosa cinnamonea,* cut in two to
show the nuts.　The leaves are
pinnate, consisting of two or
more pairs of leaflets, and ending
with an odd one.　The leaves
are furnished with very large
stipules (see *fig.* 20 *e*) ; and the
stems have numerous prickles
(*f*), which differ from thorns in
being articulated, that is, they
may be taken off without tearing the bark of
the stem on which they grow, only leaving the
scar or mark, shown at *g.*　The leaves of the
sweet briar are full of small glands or cells
filled with fragrant oil, which may be distinctly
seen in the shape of little white dots, when
held up to the light ; and this is the reason of
their delightful perfume.　When the leaf is
rubbed between the fingers, the thin skin that
covers the cells is broken, and the oil being
permitted to escape, the fragrance is increased.
There are only two genera in this tribe, viz.
*Rosa* and *Lowea,* the latter containing only
what was formerly called *Rosa berberifolia,* and
which has been thought worthy of being made
into a separate genus principally on account of
its having simple leaves without stipules, and
branched prickles.

FIG. 22.—RIPE FRUIT AND DETACHED SEED OF A ROSE.

### TRIBE II.—POTENTILLEÆ OR DRYADEÆ.

THE plants belonging to this tribe agree more
or less in the construction of their flowers with
the well-known showy plants called Potentilla,
but my readers will probably be surprised to
hear that the raspberry and the strawberry are
included among them.  If, however, they compare the flower of the Potentilla with that of
the strawberry, they will find them very much
alike.  In both there is a calyx of ten sepals,
and a cup-shaped corolla of five petals ; and in
both the stamens form a ring round an elevated receptacle, on which are placed numerous
carpels.  Here, however, the resemblance ceases,
for as the seeds of Potentilla ripen, the receptacle withers up in proportion to the swelling of
the carpels, till it becomes hidden by them;
while in the strawberry the receptacle becomes
gradually more and more dilated, swelling out
and separating the bony carpels still farther and
farther from each other, till at last it forms
what we call the ripe fruit.  I have already
had several times occasion to mention the
receptacle, which though seldom seen, or at
least noticed, by persons who are not botanists,
is a most important part of the flower, and one
that assumes a greater variety of form than any

other. Sometimes, as we have seen in several
of the Ranunculaceæ and Leguminosæ, it is a
mere disk or flat substance serving as a foundation to hold together the other parts of the
flower ; and at other times we have found it
drawn out into a thin membrane and divided
into a kind of leaves, as it is among the carpels
of the tree-peony; but in no plants that I
have yet had occasion to describe does it
assume such strange forms as in Rosaceæ.

The flower of the strawberry (*Fragaria vesca*)
has a green calyx of ten sepals; five of which
are much smaller than the others, and grow a
little behind them, the large and small ones
occurring alternately. The corolla is cup-shaped,
and in five equal petals; the stamens are
numerous and arranged in a crowded ring round
the carpels, which are placed on a somewhat
raised receptacle. The carpels or nuts resemble
those of the rose, but they have no hairy
covering, and indeed look hard and shining on
the surface of the distended receptacle, or poly-
phore as it is called in its metamorphosed state.
The carpels when ripe do not open to discharge
the seed, and consequently as they are sown with
the seeds, the young plants are a long time
before they appear. The strawberry has what
is called ternate leaves, that is, leaves consisting
of three leaflets; with large membranous stipules.

The calyx is persistent, that is, it remains on till the fruit is ripe.

The Raspberry (*Rubus Idæus*) differs widely from the strawberry in many particulars, notwithstanding their being included not only in the same natural order, but in the same tribe. The calyx has only five sepals (*a* in *figure* 23); and though the corolla has five petals (*b*), they do not form a cup-shaped flower. In the centre are the carpels, the form of which

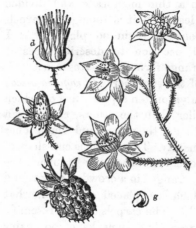

Fig. 23.—Flowers and fruit of the Raspberry.

is shown of the natural size at *c*, and magnified at *d*, the latter showing that each has a separate style and stigma. As the raspberry advances, the petals drop, and the receptacle becomes elevated into what is called a torus, as shown of the natural size at *e*; bearing the carpels upon it, which gradually swell out and soften, till each becomes a little pulpy fruit,

full of juice, and having the stone or seed in
the centre. While this change is taking place,
the stamens gradually wither and fall off, and
the stigmas disappear, the style shrivelling up
to the appearance of a hair ; the pulpy carpels
have also become so pressed against each other,
as to adhere together, and the whole, with the
persistent calyx, now assumes the appearance
shown at $f$. As soon as the carpels become
ripe they cease to adhere to the torus, and
they may be pulled off and eaten (the torus, or
core as it is called, being thrown away) : each
carpel will be found to inclose a very hard seed or
stone, as shown at $g$. If the Raspberry, instead of
being gathered, be suffered to remain on the stalk,
the juicy carpels dry up, and fall with the seed
inclosed. The stems of the Raspberry are bien-
nial, that is, they do not bear till they are two
years old, after which they die ; but the roots
are perennial, and they are always sending up
fresh suckers, so that the same plants will bear
for many years in succession, though not on the
same stems. The stems are generally erect, and
prickly like the rose ; and the leaves on the bearing stems have three leaflets, while those on the
barren stems have five ; and in both cases the
leaflets are covered with white down on the
under side. All the different kinds of Bramble,
such as the Dewberry, Blackberry, &c., agree

with the Raspberry in the construction of their
fruit, though they differ in the number of their
leaflets, the size and colour of their flowers, and
other minor particulars.

Several other genera belong to this tribe,
among which may be mentioned *Geum Avens*, or
*Herb Bennet*, the carpels of which have each a
hooked style; *Sieversia* separated from Geum,
because the carpels end in a straight feathery
awn ; and *Tormentilla*, the flowers of which bear
a general resemblance to those of Potentilla,
but which have an eight-parted calyx; a corolla
of four petals; sixteen stamens, and dry wrinkled
carpels on a depressed receptacle. All these
genera my readers will find it interesting to
procure flowers of, in order to compare them
with each other. This and the preceding tribe
are considered by some modern botanists to
form the order Rosaceæ; the other tribes being
formed into separate orders.

---

### TRIBE III.—SPIRÆEÆ.

The only genera in this tribe which contain
well-known plants are Spiræa and Kerria.
In Spiræa the calyx is five-cleft (see *a* in *fig.*
24) and lined with the dilated receptacle, forming
a shallow tube or rather cup for the reception
of the carpels. There are five small roundish

petals (*b*), and from twenty to fifty stamens (*c*),
which project very far beyond them.   In the

FIG. 24.—FLOWER OF THE SPIRÆA.

centre are from two to five carpels (*d*), which
are something like those of the raspberry when
young, but afterwards become of the kind called
follicles ; each carpel contains from two to six
seeds affixed to its inner suture, and they are
dehiscent—that is, they open naturally at the
top to discharge the seed (see *e*).   The flowers
are set very close together, and from this circumstance, combined with their small size and projecting stamens, they look like fine filigree work ;
hence the popular English names given to *S.
salicifolia* or Bridewort, Queen's needle-work,
&c.   The flowers of this species are in spicate
racemes, but others are in corymbs, as in *S. bella ;*
or in panicles, as in *S. ariæfolia.*

Kerria is a genus containing only one species,
the plant which was formerly called *Corchorus
japonica ;* the calyx is united at the base, but
divided in the upper part into five lobes ; three
of them obtuse, and the other two tipped with

a little point called a mucro.  There are about
twenty stamens about the same length as the
petals arising from the calyx, and five roundish
carpels containing one seed each.  The leaves
are simple, and the stipules awl-shaped.  Till
lately only a double-flowered variety was known
in Britain; but about 1832, the single-flowered
plant was introduced from China.  Corchorus,
the genus in which this plant was originally
placed, is nearly allied to the lime-tree.

### TRIBE IV.—AMYGDALEÆ.

This tribe is distinguished by the fruit, which
is what botanists call a drupe, that is, a stone
fruit.  The principal genera included in this
tribe are *Amygdalus*, the Almond; *Persica*, the
Peach and Nectarine; *Armeniaca*, the Apricot;
*Prunus*, the Plum; and *Cerasus*, the Cherry.
All these genera contain more or less of prussic
acid, which is found to exist principally in the
leaves and kernels; and they all yield gum when
wounded.

The flowers of the common Almond (*Amygdalus communis*) appear, as is well known, before
the leaves, bursting from large scaly buds, which
when they open throw off the brown shining
bracts in which they had been enwrapped.  The
calyx is somewhat campanulate, with the upper

part cut into five teeth or lobes, and it is lined
by the dilated disk. There are five petals, and
about twenty stamens, both inserted in the
lining of the calyx. The anthers are innate,
and they differ from most of the other plants
yet described in being only one-celled. The ovary
is also only one-celled, and there are generally
two ovules, though the plant rarely ripens more
than one seed. The leaves are simple, and
they have very small stipules. When the petals
drop, the ovary appears covered with a thick
tough downy pericarp, within which is the
hard stone or nut, the kernel or almond of which
is the seed.

The Peach (*Persica vulgaris*) was formerly
included in the same genus as the almond ; and
in fact there is but little botanical difference.
The flowers are the same both in construction
and appearance ; and the leaves are simple like
those of the almond, and, like them, they are
conduplicate (that is, folded together at the midrib) when young. The only difference indeed is in
the fruit ; for, as everybody knows, the stone
of the peach has not a dry tough covering, like
that of the almond, but a soft and melting one
full of juice, and the stone itself is of a harder
consistence, and deeply furrowed, instead of being
only slightly pitted. The fruit of the peach has
thus a fleshy pericarp, the pulp or sarcocarp of

which is eatable, and a furrowed nut or stone, inclosing the seed or kernel, which is wrapped up like that of the almond, in a thick loose skin.

The Nectarine (*P. lævis*) only differs from the peach in the epicarp, or outer covering of the pulpy part, being smooth instead of downy. Of both fruits there are two kinds, one called free-stone, from their parting freely with the stone; and the other cling-stone, from the stone clinging to the fibres of the pulp.

The Apricot (*Armeniaca vulgaris*) agrees with the preceding genus in its flowers; but it differs in its fruit, its stone being sharp at one end and blunt at the other, with a furrow on each side, but the rest of the surface smooth. Thus my readers will perceive that the Peach and the Apricot, though so different from each other as to be recognised at a glance, are yet botanically so very closely allied, as to be distinguished only by the stone. The leaves indeed differ in form, but in other respects they are exactly the same.

The Sloe (*Prunus spinosa*) is supposed by some botanists to be the origin of our cultivated plum, though others make it a separate species under the name of *Prunus domestica*. The flowers in both are solitary (see *fig.* 25), and consist of a five-toothed calyx (*a*) which is united at the base, and in the lining of which the stamens are inserted as shown at *b*. The ovary has a thick style and capitate stigma (*c*),

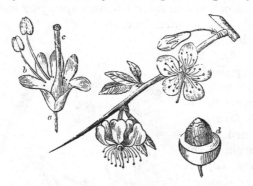

FIG. 25.—FLOWERS AND FRUIT OF THE SLOE.

and the fruit is a drupe (*d*). In these particulars therefore the plum agrees with the preceding genera ; but it will be found to differ in the skin of the pericarp, which is quite smooth and covered with a fine bloom ; this, indeed, and its stone being pointed at both ends constitute the chief botanical distinctions between the fruit of the plum and that of the apricot, as in other respects they are alike. Both the plum and the apricot have footstalks, and in this differ from the peach and the nectarine, which are without. The leaves of the plum differ from those of the other genera in being convolute, that is, rolled up, in the bud.

The Cherry (*Cerasus vulgaris*) differs from the plum in the skin of the pericarp being destitute of bloom, and in several flowers springing from each bud, in what botanists call a fascicled umbel (see *a* in *fig.* 26). The pedicels (*b*) are also much longer; the petals (*c*) are indented in the margin ; the style (*d*) is more slender ; and the stone (*e*) is smooth and much more globose. The number of the stamens, and the manner in which they are inserted in the lining of the calyx, is the same in both genera (see *f*) ; but

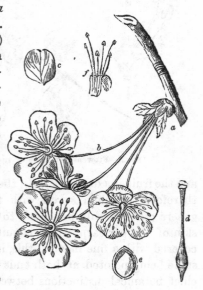

FIG. 26.—FLOWERS AND STONE OF THE CHERRY.

the leaves are different, for those of the Cherry are folded down the middle, when young, like those of the peach and almond ; while those of the plum are rolled up.

The genus Cerasus is divided into two sections, the first containing those species which have their flowers in bunches, and on long footstalks, as in the common Cherry; and the second those which have their flowers in racemes on short footstalks, as in the Bird-cherry (*Cerasus Padus*); the Mahaleb, or Bois de Sainte Lucie (*Cerasus Mahaleb*); the common Laurel (*Cerasus Lauro-Cerasus*); and the Portugal Laurel (*Cerasus lusitanicus*). These plants are so different from the common Cherry both in flowers and fruit, as far as can be judged from their general appearance, as scarcely to be recognised; but when closely examined their botanical construction will be found the same. Formerly only two genera were included in this tribe—viz. Amygdalus, which comprised the Peach and Nectarine as well as the Almond; and Prunus, which included the Apricot and the Cherry.

---

## TRIBE V.—POMEÆ.

THE common apple (*Pyrus Malus*) may be considered the type of this tribe, which comprehends not only what we are accustomed to call kerneled fruit, but also the Hawthorn, Cotoneaster, and other ornamental shrubs and low trees. The flower of the apple bears con-

siderable resemblance to the flowers of the
genera already described, but the petals (see
*a* in *fig.* 27) are oblong, rather than roundish.

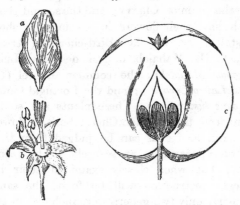

FIG. 27.—FRUIT AND PART OF THE FLOWER OF THE APPLE.

The calyx (*b*) is tubular in the lower part, and the
limb is divided into five lobes.  The receptacle
lines the lower part of the calyx, and forms a
disk, filling its throat, in which the stamens and
petals are inserted.  There are five ovaries, the
styles of which are for half their length united,
leaving the upper part and the stigmas free;
and the ovaries themselves, now become cells,
are enclosed in a cartilaginous endocarp, which
forms what we call the core of the Apple, and
which adheres firmly to the tubular part of the
calyx.  There are two ovules in each cell, placed

side by side, but generally only one seed in each becomes perfectly ripe. As the seeds advance, the fleshy tube of the calyx swells out and becomes what we call the apple ; while the leafy part or lobes of the limb remain on, and form the eye. Fruit of this kind are called pomes.

The Pear (*Pyrus communis*) differs from the apple in the shape of the fruit (see *a* in *fig.* 28),

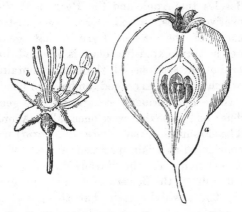

FIG. 28.—FRUIT AND PART OF THE FLOWER OF THE PEAR.

which tapers towards the footstalk, instead of being umbilicate, that is, indented at the point of the insertion of the footstalk, as is the case with the Apple. The construction of the flowers in both species is the same, except that the styles are quite free for their whole length in the Pear, and not partially united into a column

F 2

as in the Apple. This distinction, and some others,
have been thought, by some botanists, sufficient
to constitute the Apple and its allied species into
a separate genus under the name of *Malus*.
The leaves of the Pear differ from those of the
Apple in being the same colour on both surfaces,
whereas those of the Apple are covered with
a white down on the under side.

Besides the Apple and the Pear, and their
respective allies, which form two distinct
sections of the genus Pyrus, that genus,
being a very extensive one, is divided into
several other sections, all the plants contained
in which may be arranged under two heads :
viz., those that formerly constituted the genus
Sorbus ; and those that were once called Aronia.

The Mountain Ash (*Pyrus aucuparia*) may
be considered as a fair specimen of most of the
trees belonging to the Sorbus division.    By
the details of the flowers of this species given
in *fig.* 29, it will be seen that the petals (*a*)

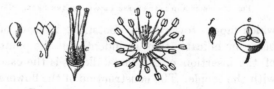

FIG. 29.—FLOWER AND FRUIT OF THE MOUNTAIN ASH.

are very small and concave ; and the calyx (*b*)

is tubular, and five-cleft. There are three
styles, as shown at *c ;* and the stamens (*d*), which
project far beyond the petals, are inserted in
the disk. The fruit (*e*) is a pome with three
seeds (*f*) enclosed in a cartilaginous membrane,
like the core of the apple or pear. The leaves
of the Mountain Ash are impari-pinnate, that
is, they consist of several pairs of leaflets, terminating in an odd one; and the flowers are
produced in corymbs. The White Beam-tree
(*Pyrus Aria*), the wild Service (*P. torminalis*),
and several similar trees, belong to this division
and have the same kind of fruit as the Mountain Ash. The true Service, however, differs in
its fruit being generally shaped like a pear,
though there is a variety with apple-shaped
fruit. One species (*P. pinnatifida*) has the
leaves lobed to the midrib, instead of being cut
into leaflets; and this gives the name to the
species, leaves of this description being called
pinnatifid. The leaves of the genus Pyrus
often have their petioles dilated and somewhat
stem-clasping at the base; but they have generally only small stipules.

Among the other plants included in the genus
Pyrus, may be mentioned the beautiful shrub
now called *Pyrus arbutifolia*, which has been
successively included in the genera Cratægus,
Aronia, and Mespilus; and *P. Chamæmespilus*,

which has been successively called Cratægus, Mespilus, and Sorbus. There are several beautiful low shrubs belonging to this division of the genus Pyrus.

The genus Cydonia, the Quince, differs from Pyrus in having its seeds arranged in longitudinal rows, instead of being placed side by side. In the Chinese Quince there are thirty seeds in each row, arranged lengthways of the fruit. The ovary of this genus consists of five cells, each containing one row of seeds, the seeds being covered with a kind of mucilaginous pulp. The well-known plant, formerly called *Pyrus japonica*, has been removed to the genus Cydonia on account of its ovary and the disposition of its seeds, which are decidedly those of the Quince. It differs, however, from the common Quince in its seeds, which are arranged in two rows in each cell.

The common Hawthorn (*Cratægus Oxyacantha*) has generally only two styles (see *a, fig.* 30), but the number of styles varies in the many different species included in the genus from one to five. The corolla, calyx, and stamens are the same as

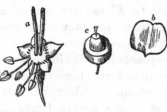

FIG. 30.—FLOWER AND FRUIT OF THE HAWTHORN.

in the other genera included in this tribe, but
the petals (*b*) are rounder and rather more indented. The seeds vary from one to five, each
being enclosed in a bony covering, or stone, the
whole being surrounded by the fleshy part of
the calyx, which forms the eatable part of the
Haw. In some of the species the haws are so
large as to appear like little apples ; but they
may be always easily distinguished by the ripe
ovary, or case which incloses the seed, being
bony ; whereas in all the varieties of Pyrus,
the outer part of the ovary is cartilaginous, like
the core of the apple. The seeds of the Hawthorn
are a long time before they come up, from the
hardness of this bony covering, which does not
open naturally when ripe. The species composing the genus Raphiolepis, the Indian Hawthorn, have been separated from Cratægus ;
chiefly on account of the covering which encloses the seeds being of a paper-like texture,
instead of bony, and each cell containing two
seeds. The limb of the calyx also falls off before
the fruit is ripe, instead of remaining on to form
what is called an eye, as it does in the common
Hawthorn. The leaves of the plants belonging
to this genus vary in the different species ; but
those of the common Hawthorn are wedge-
shaped, and cut deeply into three or five lobes.

The different species which compose the genus

Cratægus were formerly considered to belong
to the genus Mespilus. This genus, which is
now almost confined to the common Medlar
(*Mespilus germanica*), agrees with Cratægus in
having each seed enclosed in a bony covering,
but it differs in the limb of the calyx being
in large leafy segments ; and in the disk being
very large and visible even when the fruit is ripe,
from the tubular part of the calyx not closing
over it.

Among the plants formerly included in the
genus Mespilus, may be mentioned *Photinia
serrulata*, and *Eriobotrya japonica*, both natives
of Japan. The first of these was once
called *Cratægus glabra*, and it is remarkable for
its beautiful glossy leaves, which are of a deep
green when old, and beautifully tinged with red
when young ; the flowers are white, and they
are produced in what botanists call corymbose
panicles. There are some other species of the
genus Photinia, but only two or three are common in British gardens. *Eriobotrya japonica*,
the Loquat-tree, was formerly called *Mespilus
japonica*. It is remarkable for its large and
handsome leaves, which are woolly on the under
side. The flowers, which are small and white,
are produced in large panicles, and they are
followed by large pendulous bunches of the
yellow pear-shaped fruit, which is covered with

a woolly substance, and hence the botanic name
Eriobotrya, which signifies woolly grapes.
The tree will stand out in the open air in
England, and it will flower freely in a green-
house, but it requires a stove to ripen its fruit.

Cotoneaster and Amelanchier were also formerly included in Mespilus, and they are very
closely allied to Photinia and Eriobotrya.   The
species belonging to Photinia, however, are
easily known by their shining leaves, and the
petals of their flowers being reflexed, that is,
curved back ; and the species of Eriobotrya are
distinguished by their woolliness, which spreads
over even the flowers and fruit.   The Coto-
neasters are known by the small petals of their
flowers, which curve inwards, and remain a long
time without falling.   The leaves are also thick,
and woolly or clothed with rusty hair on the
under side ; and the flowers, which are produced in cymes or panicles, with woolly pedicels, are followed by bright red haws, resembling those of the hawthorn.   Lastly, the genus
Amelanchier is known by its long narrow petals,
and its ovary having five or ten cells, with five
styles united at the base.

---

### TRIBE VI.—SANGUISORBEÆ.

THE plants included in this tribe agree more
or less with the common Burnet (*Sanguisorba*

*officinalis*). This plant, which is found in great
abundance in rich meadows on calcareous soils,
has its flowers produced in a close terminal
spike. The flowers have no petals, but the
calyx, which is four-cleft, is pink, and there are
four glossy brown bracts to each flower; so that,
on the whole, the flowers are rather ornamental,
notwithstanding their want of petals. There are
only four stamens, and two carpels with slender
styles and pointed stigmas. The leaves are
pinnate, consisting generally of nine leaflets,
and each pair of leaflets is furnished with two
stipules. The Alchemilla, or Ladies' Mantle, is
nearly allied to the Burnet; but the flowers are
in small corymbs, instead of spikes. The
flowers have no petals; but the limb of the
calyx is coloured, and divided into eight unequal segments. There are generally four
stamens and only one style, though sometimes
there are two. The ovary contains one or two
carpels, each containing a single seed, and these
when ripe are enclosed in a capsule, formed by
the tubular part of the calyx becoming hardened.
The leaves are lobed, plaited, and serrated at
the margin; and those of the Alpine species
(*A. alpina*), which is often found wild on the
Scotch mountains, are covered with a beautiful silky substance of the most brilliant white-
ness.

# CHAPTER IV.

THE ORDER ONAGRACEÆ: ILLUSTRATED BY THE DIFFERENT
KINDS OF FUCHSIA; ŒNOTHERA, OR THE EVENING TREE-
PRIMROSE; GODETIA; EPILOBIUM, OR THE FRENCH WILLOW-
HERB; AND CLARKIA.

THE type of this order is considered to be the
common evening Tree-primrose (*Œnothera biennis*), and it takes its name from *Onagra*, the
name given by Tournefort to the genus. The
Fuchsia seems so unlike the Œnothera, that it
appears difficult to any but a botanist to trace
the connexion between them; but, botanically,
they agree in the position of the ovary, which
in both is so placed as to seem rather to belong
to the flower-stalk than to the flower; and this
peculiarity is found in all the genera included in
the order. The parts of the flowers are also
always either two, four, eight, or twelve; as, for
example, there are four petals and eight stamens
in both the Fuchsia and the Œnothera.

## THE GENUS FUCHSIA.

LITTLE more than fifty years ago, the first
Fuchsia was introduced into England; and we
are told that small plants of it were sold at

a guinea each.   Now more than twenty species,
and innumerable hybrids and varieties, are in
common cultivation, and we find them not only
in greenhouses and windows, but planted in the
open air as common border shrubs.   The first
Fuchsia seen in England was *F. coccinea*, introduced in 1788 ; and this species is still common
in our gardens.   It was followed about 1796 by
*F. lycoides ;* and after that no other species was
introduced till 1821, since when a full tide of
Fuchsias has kept pouring in upon our gardens,
from the different parts of Mexico, South
America, and New Zealand, to the present time.

All the Fuchsias were formerly divided into
two sections ; the plants in one of which having
the stamens and pistil concealed, and those in
the other having the stamens and style exserted,
that is, projecting beyond the other parts of the
flower.   The first division comprises all the
small-flowered kinds ; such as *F. microphylla,
thymifolia, cylindracea,* and *bacillaris,* all which
have the lobes of the calyx short, and the petals
partially concealed.   *F. parviflora* belongs to
this division, but it is distinguished by its glaucous leaves with an entire margin ; and *F. lycoides*
is also included in it ; though this last seems to
form the connecting link between the two
sections, as both its petals and its style and stamens are partially exposed.   The second division

comprises all the kinds which have long projecting stamens.

As the general arrangement of the parts of the flower is nearly the same in both divisions, *fig.* 31, which represents the section of a flower of *F. cylindracea*, from the *Botanical Register*, will give my readers a clear idea of the botanical construction of the Fuchsia. In this figure, *a* shows two cells of the ovary (which when entire is four-celled, opening when ripe into four valves), with the seeds attached to a central placenta. This ovary is surrounded and protected by the dilated disk, which also serves as a lining to the tubular part of the calyx, *b*. The anthers, in this division, have very short filaments, which are inserted in the lining of the calyx, as shown at *c* ; *d* is the style, which, in fact, consists of four styles united together, and which divides near the apex into four stigmas ; *e e* are two of the four lobes of the calyx ; and *f* is one of the four petals.

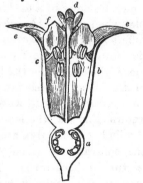

FIG. 31.—SECTION OF THE FLOWER OF FUCHSIA CYLINDRACEA.

In the second division, of which *F. coccinea*

may be considered the type, the calyx and the corolla are of different colours. In *fig.* 32, which shows a flower of *F. discolor*, the Port Famine Fuchsia, the calyx (*a*) is scarlet and the most ornamental part of the flower, while the petals (*b*) are purple, and wrapped over each other. The ovary (*c*) is green,

and when the petals and calyx fall off, it swells into a berry, which becomes of a dark purple when ripe. *F. globosa* differs from *F. coccinea* in the flowers being shorter and more globose, while the limb of the calyx curves inward. In *F. macrostemma*, a well-known Fuchsia, the lobes of the limb of the calyx are, on the contrary, recurved, that is, turned backwards. This

Fig. 32.—Fuchsia Discolor.

formation is common, more or less, to several other species. In *F. excorticata*, the New Zealand Fuchsia, there is a large fleshy knot at the base of the calyx, and strong ribs running up the lobes; the calyx is green when young, but it afterwards becomes crimson ; and the petals are very small. This species is so different from the others, that it was at first described as a new genus, under the name of Skinnera. The calyx is green at first, but it afterwards becomes crimson. *F. arborea* has pale-purplish flowers, and, like *F. lycoides*, forms a connecting link between the two sections, the stamens being only a little exserted, and the petals hidden.

*F. radicans*, the only Fuchsia yet discovered with a creeping stem, which was introduced in 1841, belongs to this division.

These sections include all the Fuchsias known in British gardens previously to 1835 ; but since that period, two kinds have been introduced, which belong to a third division. These are *F. fulgens* and *F. corymbiflora*. In these plants the tube of the calyx is about two inches long, and the lobes are very short. The petals are also short, and scarlet or deep-rose colour, though not exactly of the same hue as the calyx. The leaves are large, with the midribs and veins red ; and the branches and pedicels are also of a dark reddish purple.

### THE GENUS ŒNOTHERA.

In the description of the botanical construction of the Fuchsia, my readers may have observed, that the ovary is placed below the calyx, and quite distinct from it. The same construction is still more visible in the Œnothera, as the tube of the calyx is very slender, and often more than two inches long, while the ovary is often vase-shaped, and of large size. The calyx of *Œno-thera biennis*, the common Evening or Tree Primrose, consists of four sepals growing together in the lower part, so as to form a long tube (*a* in *fig.* 33), and with the upper part or limb gene-rally in two segments (*b*), which are bent quite

FIG. 33.—THE EVENING PRIMROSE (*Œnothera biennis*).

back when the corolla expands, and which may be easily divided with a pin into four. There are four petals in the corolla (*c*), and they are placed so as to wrap over each other at the base. The calyx is lined with the dilated receptacle, and in this lining are inserted the filaments of the eight stamens (as shown at *d*); the stamens

CHAP. IV.]     THE GENUS ŒNOTHERA.     81

having versatile anthers, that is, anthers attached
to the filament by the middle, so as to quiver
at every breath. The pollen contained in the
cells of these anthers feels clammy when touched;
and its particles, when magnified, will be found
to be triangular, and connected by small threads,
a form of construction peculiar to this genus
and its allies. The style is long, and the stigma
is four-cleft. The ovary (*e e*) is situated at the
base of the calyx, and when ripe, it becomes a
four-celled dry capsule, which bursts into four
valves, opening at top to discharge the seed.
The seeds, when young, are attached to the
central placenta, and they are quite free from
hair or wool of any kind.

The genus Œnothera being a very extensive
one, it has been divided by M. Spach, a German
botanist residing in Paris, into fourteen new
genera ; but only one, or at most two, of these
genera have been adopted by other botanists.
One of these Godetia, which embraces all the
purple-flowered kinds, has been divided from
Œnothera, on account of a slight feathery appearance on the seeds ; whereas the seeds of
the true yellow-flowered Œnotheras are naked,
that is, without the slightest appearance of any
feathery substance or wing. The other genus,
Boisduvalia *Spach*, includes only two species,
both with pink flowers, which are very seldom

seen in British gardens. The generic mark of
distinction consists in four of the stamens in
these species being shorter than the other four ;
whereas in the true Œnotheras all the eight
stamens are of equal length. As M. Spach's
other genera have not been adopted by any
British botanist, it is not worth while troubling
my readers with the distinctions between them.
The flowers of the yellow Œnotheras only open
in the evening, or in cloudy weather ; but
those of the purple kinds, or Godetias, remain
open all day. The leaves in both kinds are
alternate.

### THE GENUS EPILOBIUM.

THIS genus is well known, by the showy
plant often seen
in shrubberies,
called the French
Willow - Herb —
(*Epilobium an-
gustifolium*), and
the English weed
called Codlings-
and - Cream (*E.
hirsutum*). In this
genus, the tubular
part of the calyx

FIG. 34.—EPILOBIUM ROSEUM.

which incloses the ovary, is quadrangular, as

shown at *a* in *fig.* 34, which represents seed-
vessels of *Epilobium roseum,* a very common
weed in the neighbourhood of London. The
limb of the calyx is four-cleft, and the corolla
has four petals ; and when these fall off, the
ovary assumes the appearance shown
at *a.* The quadrangular form is retained by the capsule, which, when
it ripens, bursts open into the four
valves (*b*), and discharges the seed
which was attached to the central

FIG. 35.—SEED
OF EPILOBIUM.

placenta (*c*) ; each seed being furnished with a
little feathery tuft resembling pappus, as shown
in *fig.* 35. The genus Epilobium is divided into
two sections ; the plants in one of which have
irregular petals, the stamens bent, and the
stigma divided into four lobes, as in the French
Willow-Herb, and the other showy species ; and
the plants in the other section having small
flowers with regular petals, erect stamens, and
the stigma undivided.

### THE GENUS CLARKIA.

THE calyx in this genus is tubular, with the
limb in two or four lobes, as in Œnothera. The
corolla is, however, very different, the four petals
being unguiculate or clawed ; that is, so much
narrower in the lower part as to stand widely
apart from each other ; they are also three

lobed. The stamens are very different, only four of them being perfect, and the anthers of the other four being wasted and destitute of pollen; and the stigma is divided into four leaf-like lobes, very different from those of all the other genera included in the order. The capsule is cylindrical in shape, and furrowed on the outside; it is four-celled, and when ripe, it bursts open by four valves. The seeds are quite naked.

Among the other genera belonging to this order, I may mention the following: *Gaura,* the petals of which are somewhat unguiculate, like those of Clarkia, but not three-lobed as in that genus; the segments of the limb of the calyx often adhere two together, so as to appear three instead of four; the ovary is one-celled, and the seeds naked: *Lopezia,* which has apparently five irregular petals, though, on examination, one will be found to be a meta-morphosed stamen, a four-cleft calyx, two stamens, including the one converted into a petal, and a globular, four-celled capsule: and *Circæa,* or Enchanter's Nightshade, which has the limb of the calyx apparently in only two segments, and only two petals and two stamens; the capsule is globular like that of *Lopezia,* but it is covered with very small hooked bristles, and it is divided into only two cells, each containing only one seed.

## CHAPTER V.

THE ORDER RUBIACEÆ: ILLUSTRATED BY THE CINCHONA, OR
PERUVIAN BARK; LUCULIA GRATISSIMA; CAPE JASMINE;
RONDELETIA; COFFEE; IXORA; IPECACUANHA; MADDER;
GALIUM; WOODRUFF; AND CRUCINELLA STYLOSA.

THIS order contains more than two hundred
genera; but by far the greater part of these are
composed of tropical plants, many of which are
not yet introduced into Britain.   Several of the
genera, on the other hand, are British weeds;
and this difference in habit, with others in the
qualities of the plants, &c., have occasioned
some botanists to divide the order into two :  one
of the new orders being called Cinchonaceæ, and
containing the plants most resembling Cinchona;
and the other Galiaceæ, containing the plants
most nearly allied to Galium or Bedstraw.

The characteristics of Rubiaceæ, in its most
extended sense, are that the ovary is surrounded
by the calyx, and placed below the rest of the
flower; and that the corolla has a long tube, lined
with the dilated receptacle, in which the stamens
are inserted.   In most of the species, the filaments are very short, and the anthers nearly or
entirely hidden in the corolla; and in many cases,

the segments of the calyx remain on the ripe
fruit, as they do in the genus Pyrus in Rosaceæ,
where they form what is called the eye in the
apple and pear.

The qualities of the Cinchona division of the
Rubiaceæ are generally tonic ; but some of the
plants, as for example the Ipecacuanha, are
used as emetics, and one (*Randia dumetorum*) is
poisonous. The qualities of the Galium division
are not so decidedly marked ; but the roots of
some of the plants are used for dyeing.

### THE GENUS CINCHONA, AND ITS ALLIES.

THE well-known medicine called Peruvian
bark is produced by three species of the genus
Cinchona ; the pale bark, which is considered
the best, being that of *C. lanceolata*. The flowers
of this species are small, and of a very pale pink.
The calyx (see *a* in fig. 36) is bell-shaped, and
five-toothed ; and the corolla (*b*) is tubular, with
the limb divided into five lobes, and silky within,
as shown in the magnified section at *c*. The
stamens (*d*) have very short filaments, which
are inserted in the throat of the corolla. The
ovary (*e*), which is deeply furrowed when young,
is inclosed in the calyx ; it is two-celled, with a
single style, and a two-lobed stigma (*f*). The
capsules retain the lobes of the calyx as a sort
of crown (*g*); and they open naturally at the

division between the two cells, as shown at *h*,
beginning at the base.    The cells (*i*) each con-

FIG. 36.—CINCHONA, PERUVIAN BARK (CINCHONA LANCEOLATA).

tain several seeds.    *C. oblongifolia*, which yields
the red bark of the shops, has cream-coloured
flowers, as large as those of a Jasmine, which
they resemble in shape ; and *C. cordifolia*, which
produces the yellow bark, has flowers like the
first species, and heart-shaped leaves.    The singular plant called *Hillia longiflora*, is nearly
allied to Cinchona ; as is also the beautiful and
delightfully fragrant *Luculia gratissima*.    In this
last plant the tube of the calyx is very short,
and pear-shaped, and the segments of the limb
are short, and sharply pointed.    The corolla is
salver-shaped, with a long tube, and a spreading,
five-parted limb. The anthers are nearly sessile,
and the short filaments to which they are at-

tached are inserted in the throat of the corolla,
only the tips of the anthers being visible. The
stigma is divided into two fleshy lobes, and the
capsule splits, not like that of Cinchona, but
from the apex to the base in the centre of each
cell. The seeds are very small, and each has a
toothed, membranous wing. The flowers of
this beautiful plant are produced in a large head,
and at first sight greatly resemble those of a
Hydrangea; but they are easily distinguished by
their delightful fragrance.

*Manettia cordifolia*, a very pretty stove-twiner
often seen in collections, is very nearly allied to
Luculia, differing principally in the shape of the
flowers, which in Manettia have a long tube
and a very small limb. *Bouvardia triphylla* and
the other species of Bouvardia, and *Pinckneya
pubescens*, belong to this division; and such of
my readers as have the living plants to refer to,
will find it both interesting and instructive to
dissect them and compare the parts of their
flowers with the description I have given of
Luculia and Cinchona, so as to discover the
difference between the different genera; afterwards reading the generic character of each
given in botanical works, that they may see how
far they were right.

### THE GENUS GARDENIA AND ITS ALLIES.

THE Cape Jasmine (*Gardenia radicans*) is a well-known greenhouse plant, remarkable for the heavy fragrance of its large white flowers, which die off a pale yellow, or buff. The calyx has a ribbed tube, and the limb is parted into long awl-shaped segments. The corolla is salver-shaped, that is, it has a long tube and a spreading limb, the limb being twisted in the bud. There are from five to nine anthers, having very short filaments which are inserted in the throat of the corolla. The stigma is divided into two erect fleshy lobes. The ovary is one-celled, but there are some traces of membranes, which would, if perfect, have divided it into from two to five cells. The seeds are numerous and very small. *Gardenia radicans* is a dwarf plant, which flowers freely when of very small size, and is easily propagated from the readiness with which its stem throws out roots; but *G. florida* is a shrub five or six feet high, and much more difficult to cultivate. In both species the flowers are generally double, and the petals are of a fleshy substance, which gives the corolla a peculiarly wax-like appearance.

There are many other species, but the two above-mentioned are the most common in British gardens. *Burchellia capensis* is gene-

rally considered to belong to this division of
Rubiaceæ, though its flowers bear more resemblance to those of Cinchona ; and the singular
plant called *Mussæuda pubescens*, the flowers of
which are small and yellow, but the bracts are
so large and so brilliantly white as to look like
flowers ; *Posoqueria versicolor*, an ornamental
plant lately introduced, belong to this division.

### THE GENUS RONDELETIA AND ITS ALLIES.

*Rondeletia odorata*, sometimes called *R.
coccinea*, and sometimes *R. speciosa*, is a very
fragrant stove shrub, a native of Cuba. The
flowers are produced in corymbs, and their

botanical construction is
shown in the magnified
section *fig*. 37. In this *a*
is the ovary inclosed in a
hairy calyx ; *b* shows the
limb of the calyx cut into
awl-shaped segments ; *c*
shows the manner in which
the very short filaments of
the anthers are inserted in
the throat of the corolla ;
*d* shows the termination of
the dilated receptacle which
lines the tube of the corolla ;
and *e* the segments of the

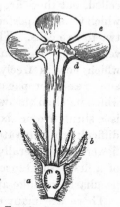

FIG. 37.—SECTION OF THE
FLOWER OF RONDELETIA.

limb. I have given the section of this flower, that my readers may compare it with the section of the flower of the Cinchona in *fig.* 36, in p. 87, and may see the general resemblance which connects the two plants in the same order, and the differences which mark them to be of different genera. *Fig.* 38 is a tuft of flowers of *Rondeletia odorata. Wendlandia* is nearly allied to Rondeletia ; as is the magnificent *Portlandia grandiflora,* which somewhat resembles *Brugmansia lutea* in shape though not in colour, as its flowers are white.

FIG. 38.—PART OF THE HEAD OF FLOWERS OF RONDELETIA.

### THE GENUS COFFEA AND ITS ALLIES.

THE Coffee-tree (*Coffea arabica*) differs from the other Rubiaceæ in the tube of its calyx being very short and disappearing when the ovary begins to swell ; and in the filaments of the stamens being sufficiently long to allow the anthers to be seen above the throat of the corolla (see *a* in *fig.* 39). The limb of the corolla (*b*) is five-cleft, and the style (*c*) bifid. Each ovary when its flower falls, becomes distended into a berry (*d*) or rather drupe, containing the nut *e*, in which are two seeds, flat

on one side, and convex on the other, which are
placed with the flat sides together, as shown at
*f*; each seed having a deep longitudinal groove,
as shown at *g*.    These seeds are our coffee.

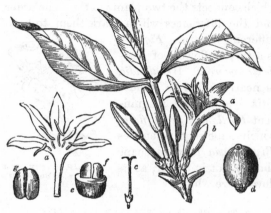

FIG. 39.—COFFEE.    (*Coffea Arabica*.)

The flowers of *Ixora coccinea* have the same
general construction as those of the other plants
of the order.    The calyx has an ovate tube, and
a very small four-toothed limb ;  and the corolla
is salver-shaped, with a long and very slender
tube, and a four-parted spreading limb.    There
are four anthers inserted in the throat of the
tube of the corolla, and just appearing beyond
it, and rising a little above them is the point of
the style with its two-cleft stigma.    The berry

is two-celled, but it differs from that of the coffee in retaining the lobes of the calyx, which form a sort of crown. There are many kinds of Ixora, all stove shrubs, and all conspicuous for their large heads or rather corymbs of showy flowers. The genus Pavetta has been divided from Ixora, principally because the species composing it have the style projecting considerably beyond the corolla, instead of only just appearing above it.

The drug called Ipecacuanha is the produce of two plants belonging to this order, *Cephælis Ipecacuanha* and *Richardsonia scabra;* though a spurious kind is made from the roots of three species of Viola, all natives of South America, and a still inferior one from the roots of a kind of Euphorbia, a native of Virginia and Carolina. It is important to know this, as the best kinds possess tonic properties as well as emetic ones, while the inferior kinds are only emetics, and they are very injurious if taken frequently. The best brown Ipecacuanha is the powdered root of *Cephælis Ipecacuanha;* a plant with small white flowers collected into a globose head, which is shrouded in an involucre closely resembling a common calyx. The true calyx to each separate flower is small and roundish, with a very short five-toothed limb. The corolla is funnel-shaped, with five small bluntish lobes.

The anthers are inclosed in the corolla, and
the stigma, which is two-cleft, projects only a
little beyond them.  The berries are two-celled
and two-seeded, and they retain the lobes of
the calyx.  The root is fleshy and creeping.
*Richardsonia scabra*, which produces the white
Ipecacuanha, has its flowers also in heads, but
the calyx is larger in proportion to the corolla,
and the stamens and style are both visible.
The capsule contains three or four one-seeded
nuts, crowned by the calyx; which, however,
becomes loosened at the base, and falls off,
before the seeds are quite ripe.  Cephalanthus,
Spermacoce, and Crusea, are nearly allied to
Richardsonia.

The above plants all agree, more or less,
with Cinchona, in their qualities, and they
are all included by Dr. Lindley in the order
Cinchonaceæ.

### THE GENUS GALIUM AND ITS ALLIES.

THE common Bedstraw (*Galium vernum*) is
a British weed, common in dry fields and on
little knolls, which produces its cluster of bright
yellow flowers in July and August. The flowers
are so small that it is difficult to examine
them in detail, but, by the aid of a microscope,
the ovary will be found to be inclosed in the
tube of the calyx as in the other Rubiaceæ,

though the calyx has hardly any limb. The corolla is what is called rotate or wheel-shaped, and its limb is divided into four segments. There are four short stamens, with their filaments inserted in the throat of the corolla, and two very short styles. The fruit is a dry capsule inclosing two seeds. Thus far the construction of the plant agrees with the other Rubiaceæ, but the stem is square, and the leaves are different, for they are without footstalks, and are disposed in what is called a whorl (see *fig*. 40). The whorl, however, according to Professor De Candolle, does not consist entirely of leaves; but of two opposite leaves and two or more stipules, which are so like the leaves as scarcely to be distinguished from them, though upon close examination,

FIG. 40.—WHORL OF LEAVES OF BEDSTRAW. (*Galium vernum.*)

it will be found that the leaves have buds in their axils (that is between them and the stem), which the stipules have not. This theory is not adopted by Dr. Lindley, who considers the whorl to consist entirely of leaves, and to be one of the distinctive marks of his order Galiaceæ.

All the plants in this division of Rubiaceæ

agree with the common Bedstraw (*Galium vernum*) in the formation of their leaves and stem ;
but the species of Galium are distinguished by
the margins of the leaves and the principal veins,
in nearly all the species, being covered with
prickles, which in some cases point forwards,
and in others are bent back, so as to catch
everything they touch.   This is particularly the
case with the leaves of the plant called Goose-
grass, or Cleavers (*Galium aparine*) ; and its
fruit is covered with hooked bristles, which take
so firm a hold as to make it difficult to separate
them from anything they have caught hold of.
The pretty little weed called Field Madder
(*Sherardia arvensis*), the fragrant Woodruff,
(*Asperula odorata*), and *Rubia peregrina*, the only
British species of Madder, all agree with Galium
in its more important characters; and as they
are all common weeds, my readers will probably
find it interesting to trace the differences
between them.   Galium and Rubia agree in
having scarcely any limb to the calyx, and a
rotate corolla; but the limb, which is only four-
parted, or even three-parted, in Galium, has always
five lobes in Rubia ; there are also five stamens
in Rubia, and the fruit is a berry; whereas
there are only four stamens in Galium, and the
fruit is dry.   Sherardia agrees with Asperula
in having a funnel-shaped corolla with a four-

cleft limb ; but in Sherardia the limb of the
calyx remains on as a crown to the fruit, while
in Asperula it drops off.   In Sherardia there is
only one style with a two-lobed stigma ;  and in
Asperula there are two styles united at the
base.

There is a very pretty plant called *Crucinella
stylosa,* which has lately been much cultivated in
gardens, and which belongs to this order.   This
plant has large heads of pretty pink flowers,
each of which has a funnel-shaped corolla, with
a long tube concealing the anthers, but beyond
which the style projects so far as to give rise to
the specific name of *stylosa.*   The stigma in this
plant is clavate, that is, club-shaped, and it is
cleft in two, though the lobes are not spreading.

H

# CHAPTER VI.

THE ORDER COMPOSITÆ: ILLUSTRATED BY THE SUCCORY, THE
SOW-THISTLE, THE DANDELION, THE BURDOCK, THE DAISY,
THE CHRYSANTHEMUM, FEVERFEW, PELLITORY OF SPAIN, WILD
CHAMOMILE, TRUE CHAMOMILE, YARROW, THE BUR-MARIGOLD,
GROUNDSEL, RAGWORT, BIRD'S TONGUE, PURPLE JACOBÆA,
CINERARIA, SUNFLOWER, MUTISIA, AND TRIPTILION.

THE plants composing the order Compositæ
have all compound flowers, which differ from
other flowers as much as a compound leaf does
from a simple one. As the compound leaf is
composed of a number of leaflets or pinnæ united
by a common petiole ; so a compound flower is
composed of a number of florets, united by a
common receptacle, which is surrounded by a
calyx-like involucre, so as to give the whole
mass the appearance of a simple flower. Each
floret has a calyx, the tubular part of which is
rarely sufficiently distinct to be perceptible, but
the limb is generally cut into long feathery
segments called pappus. The ovary of each
floret contains only one seed ; and the fruit,
which is called an achenium, retains the pappus
when ripe, and falls without opening. There
are five stamens, the filaments of which are distinct, but the anthers grow together so as to

form a kind of cylinder, through which passes
the style, ending in a two-lobed stigma (see *a*
in *fig.* 41). Most of the corollas are of two

kinds: viz. the ligulate, as exemplified in the floret of the wild
Lettuce (*Lactuca virosa*) shown in
*fig.* 41; and the tubular, as shown
in a floret of the Cotton-thistle
(*Onopordium Acanthium*) see *fig.*
42. All the British species of
Compositæ have their florets either
entirely of one of these kinds, or

FIG. 41.—LIGULATE
FLORET OF WILD
LETTUCE.

of the two mixed together; but some foreign
genera have florets with two equal lips, cut
into three or four lobes, as shown
in a floret of *Mutisia latifolia*,
at *e*, *fig.* 46, p. 108. These florets are called bilabiate. It will
be observed that in all these examples, as indeed, in all the
flowers belonging to the order,
that the pappus (*b*, in *figs.* 41 and
42), is always on the outside of
the corolla, thus plainly indicating its connexion with the
calyx.

FIG. 42.—TUBULAR
FLORET OF THE
COTTON-THISTLE.

The order Compositæ is a very large one,
above seven thousand species having been

named and described; and to assist the memory
in retaining the names of this great number of
plants, various means have been devised for
dividing the order into sections and tribes.
The principal botanists who have proposed
means of arranging this order, are Cassini,
Lessing, and lastly the late Professor De Candolle, in three volumes of his *Prodromus* published in 1840.   But as the distinctions between
the divisions proposed, lie in the difference
found in the stigmas and pappus of the different
genera, I have judged them too troublesome for
my readers, as I am sure they are for myself,
and I have preferred following the plan adopted
by Dr. Lindley in his *Elements of Botany*, published in 1841, and dividing the Compositæ into
four tribes ; viz., the three originally proposed
by Jussieu, and a fourth added by Professor
De Candolle, containing the plants with bilabiate florets, which were either not known, or
overlooked, by Jussieu.   It may perhaps be
necessary to add, that this arrangement forms
the basis of the new one proposed by De Candolle, and that the principal difference consists
in the subdivisions.

### TRIBE I.—CICHORACEÆ.

*Florets ligulate.　Juice milky, narcotic.*

THE plants contained in this tribe bear more
or less resemblance to the common Succory
(*Cichorium Intybus*).　This beautiful plant,
which is found in great abundance wild in many
of the sandy and chalky districts of England,
has large bright blue flowers, which when examined will be found to consist of a number of
florets, all of the kind called ligulate, that is
somewhat like a cornet of paper ; the upper
part being broad and flat, and serrated at the
edge.　The pappus in this genus is very short,
and it is scaly rather than feathery.　The leaves
are bitter, and when broken give out a milky
juice; and the fleshy roots when roasted are
used to adulterate coffee.　The Endive is a
variety of this species, or another species of the
same genus.　The Sowthistle (*Sonchus oleraceus*)
abounds in the same milky juice as the succory,
and has the same kind of fleshy root.　The
flower is composed of a scaly involucre (shown
at *a* in *fig.* 43) and a number of ligulate florets
(see *b*), which when they fall show the pappus
(*c*), forming a feathery ball.　The manner in
which the pappus is attached to the seed-vessel is
shown at *d ;* and the receptacle after the florets
have been pulled out, but with the involucre still

attached to it, at *e*.    A detached floret is shown
at *f*.   The Dandelion (*Leontodon Taraxacum*)

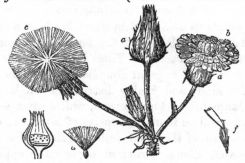

FIG. 43.—SOWTHISTLE.   (*Sonchus oleraceus.*)

differs from the Sowthistle: in its florets, which
are flatter and looser ; in its receptacle, which
is globular; and above all, in its pappus, which
is what is called stipitate or stalked, that is, the
tubular part of its calyx rises to a considerable
height above the capsule, before it becomes
divided into its feathery segments, as shown in
*fig.* 44.   The leaves of this plant are
what is called runcinate, that is, the
lobes into which they are cut point
downwards towards the root instead of
upwards from it, and the root is also
fleshy.   The Lettuce, Salsafy or Goat's-
beard, Ox-tongue, Hawkweed, Cat's-
ear, Nipplewort or Swine's Succory, and
many other well-known plants, belong
to this tribe.

FIG. 44.—
SEED OF THE
DANDELION.

TRIBE II.—CYNAROCEPHALÆ.

*Florets tubular.   Juice watery, tonic.*

THE plants in this division all bear more or less relation to the common Artichoke (*Cynara Scolymus*). The scales of the involucre are generally fleshy at the base, but terminate outwardly in a sharp hard point. The florets are tubular, and intermixed with them in the receptacle are frequently found the hardened bracts, which in this state are called paleæ, and which appear to be of a chaffy substance, as exemplified in the choke of the Artichoke, the fleshy receptacle being in this plant what we call the Artichoke bottom. This peculiar formation is shown more in detail in *fig.* 45, which represents part of the flower of the common Bur or Burdock (*Arctium Lappa*), so annoying from the strong hold it takes of any part of the dress which it may chance to touch. In *fig.* 45 *a* is the involucre, every scale in which is hooked and

FIG. 45.—PART OF THE FLOWER-HEAD OF THE BURDOCK.

turned inwards, so as to hold firmly whatever it may catch ; *b* is a floret showing its tubular shape, and its style proceeding through the

united anthers ; *c* shows the hardened bracts
or paleæ, the other florets having been removed;
and *d* shows a fruit with a palea attached, magnified.   All the different kinds of thistle belong
to this division ; and though many of the kinds
have not the hardened bracts, they have all a
spiny involucre.   The pappus of the thistle is
generally attached to a kind of disk, from which
it becomes loosened soon after the seed falls,
and this thistle, down, as it is called, being extremely light, is blown about by the winds.
All the thistles have fleshy roots, and take firm
hold of the soil.   The Corn Blue-bottles (*Cen-
taurea*), the Wild Saffron (*Carthamus tinctoria*),
and many other well-known plants, belong to
this division.

### TRIBE III.—CORYMBIFERÆ.

*Florets partly tubular and partly ligulate ; juice watery ;
sometimes bitter and tonic, and sometimes acrid.   The seeds
of some of the species yield oil.*

THE plants included in this tribe all bear
more or less resemblance to the common Daisy.
In this well-known flower, the white florets are
all ligulate, and compose what is called the ray,
and the yellow flowers, which are tubular, are
called the disk.   The involucre is simple and
leafy, and the receptacle is conical.   The seeds
are without pappus.   The Chrysanthemum is

nearly allied to the Daisy, and its seeds also
are destitute of pappus; but it is easily distinguished by its involucre, which is scaly, and by
the flower forming a kind of depressed globe in
the bud.   The scales of the involucre are
strongly marked, from being edged with a thin
membrane, and the florets of the ray are much
longer in proportion to those of the disk than
in the Daisy.   The great Ox-eye Daisy, which
was formerly called *Chrysanthemum leucanthemum*,
is now placed in a new genus, and called *Leucanthemum vulgare;* and the Chinese Chrysan-
themums have been removed to the genus *Pyrethrum*.   Both plants, however, will no doubt long
continue to be called by their old names.   The
beautiful yellow-flowered plant often found growing among corn (*Chrysanthemum segetum*), the
three-coloured Chrysanthemum (*C. tricolor* or
*carinatum*), and the yellow annual Chrysanthemum (*C. coronarium*), with some others, have
been left by Professor De Candolle in their old
genus.   In the Feverfew (*Pyrethrum*), the receptacle is elevated, and the fruit is crowned
with a narrow membrane.   The Pellitory of
Spain was formerly considered to belong to this
genus, and afterwards to the Chamomile, but it
is now called *Anacyclus Pyrethrum*.   *Matricaria
Chamomilla*, the wild Chamomile, has also no
pappus; and in this plant the receptacle is

almost cylindrical. The true Chamomile (*Anthemis nobilis*) greatly resembles the Chrysanthemum in its flowers; but they are distinguished by having a chaffy receptacle, and the fruit having a membranous margin. The smell of the Chamomile is aromatic, and its qualities highly tonic. The Yarrow (*Achillea millefolium*) is another plant destitute of pappus, but with a chaffy receptacle; it is also remarkable for its leaves, which are doubly pinnatifid.

It will be seen by the above enumeration, that in many plants belonging to this division, the pappus is entirely wanting, and in others it will be found to assume a different form to that which it bears in the other tribes. Thus, in the Bur-Marigold (*Bidens*), the pappus consists of from two to five erect awns, which are covered with very small, bent bristles. The genus Senecio has soft, hairy pappus, as may be seen in the common Groundsel (*S. vulgaris*); the leaves of this weed are pinnatifid, and somewhat stem-clasping, and the flowers have no ray florets. In other species of this division, however, the ray florets are very conspicuous: as, for example, in the common yellow Ragwort (*S. Jacobæa*), in the great fen Ragwort, or Bird's tongue (*S. paludosa*), and in the purple Jacobæa (*S. elegans*). Nearly allied to Senecio, is the genus Cineraria, so much, indeed, that Professor De Candolle,

in his late arrangement of the Compositæ, has
included the greater part of the species in Senecio. The green-house species, with purple
flowers, are among those which have been
changed; but they will probably always retain
the appellation of Cineraria, as an arbitrary
English name. The Asters, or Michaelmas
Daisies, Golden Rod, Elecampane, Leopard's
Bane, the Cape Marigold, (now called Dimor-
phortheca, instead of being included in the genus
Calendula), Coltsfoot, Wormwood, Southern-
wood, Tansy, and many other well-known plants,
belong to this division.

The Sun-flower (*Helianthemum annuus*) is an
example of one of the plants belonging to this
division which has seeds yielding oil. In this
plant the pappus is awl-shaped, and deciduous;
and the receptacle, which is broad and somewhat convex, is paleaceous. The seeds are large
and oblong, and when pressed, yield a considerable quantity of oil. The Madia is another oil
plant; and indeed the seeds of several in this
division yield oil.

---

TRIBE IV.—LABIATÆFLORÆ.
*Florets bilabiate.*

THE plants belonging to this division are
rarely seen in British gardens; but when they
do occur, they are well worth examining, from

the singularity of their formation. *Mutisia latifolia*
(see fig. 46) has a large, woolly involucre, the

scales of which are
of two kinds, the
outer ones, (*a*),
being pointed and
leaf-like, and the
inner ones, (*b*),
having the ap-
pearance of scaly
bracts. The flo-
rets of the ray,
(*c*), are narrow,
and spreading in
the fully expand-
ed flower; and
those of the disk,
(*d*), are shorter,
erect, divided into

Fig. 46.—Flower and Leaf of
Mutisia Latifolia.

two lips, which curl back, and the lower one
of which is again divided into two segments
(as shown at *e* in the detached floret). The
leaves of this plant are very curious; the midrib is lengthened and drawn out into a tendril, as shown at *f*, and the petiole (*g*) is decurrent. There are several other genera belonging
to this tribe, but none of them are particularly
ornamental except *Triptilion spinosum*, which
has flowers of the most brilliant blue, that do
not lose the intensity of their colour in drying.

## CHAPTER VII.

THE ORDER ERICACEÆ : ILLUSTRATED BY THE COMMON OR
BESOM HEATH, THE MOOR HEATH, CAPE HEATHS, LING OR
HEATHER, ANDROMEDA, LYONIA, ST. DABÆOC'S HEATH, ARBU-
TUS, THE BEARBERRY, GAULTHERIA, CLETHRA, RHODODEN-
DRON, INDIAN OR CHINESE AZALEAS, YELLOW AZALEA, AME-
RICAN AZALEAS, RHODORAK, ALMIA, MENZIESIA, LOISELEURIA,
LEDUM, LEIOPHYLLUM, THE BILBERRY, THE WHORTLE-
BERRY, THE CRANBERRY, PYROLA, AND MONOTROPA.

THE name of Ericaceæ, which most people
are aware signifies the Heath family, conjures
up immediately the image of a number of
narrow-leaved plants, with globular, ventricose,
or bell-shaped flowers ; and we are apt at first
to think that the family is so natural a one, as
to require very little explanation. Did the
order include only the Heaths, this would be the
case, for all the Heaths, different as they are in
some particulars, may be recognised at a glance:
but as the order includes the Rhododendrons,
Azaleas, and Kalmias, besides several other
plants which have not so strong a family likeness to each other as the Heaths, it becomes
necessary to say a few words on the botanical
resemblances which connect them together. The
first, and most striking, of these is the shape of
the anthers, each of which appears like two

anthers stuck together, and the manner of their
opening, which is always by a pore or round
hole, in the upper extremity of each cell.   The
filaments, also, in all the genera, except Vaccinum
and Oxycoccus, grow from beneath the seed-
vessel, being generally slightly attached to the
base of the corolla.   There is always a single
style with an undivided stigma, though the capsule has generally four cells, each containing
several of the seeds, which are small and numerous.   The calyx is four or five cleft, and the
corolla is tubular, with a larger or smaller limb,
which is also four or five cleft.   The order has
been divided into four tribes, which I shall
describe in this chapter, though some of these
are considered as separate orders by Dr. Lindley and other botanists.

---

### TRIBE I.—ERICEÆ.

THIS tribe, which comprehends all the heath-
like plants, has
been re-divided
into   two   sub -
tribes, one con-
taining   the   ge-
nera most nearly
allied   to   the
heaths, and the
other those be-

FIG. 47.—THE BESOM HEATH (*Erica Tetralix*).

longing to the Andromeda. In both there is
a honey-bearing disk under the ovary, and the
leaves are generally rolled in at the margin, as
shown at *a*, in *fig.* 47.

---

### SUB-TRIBE I.—ERICEÆ NORMALES.

ALL the genera in this sub-tribe, twenty-two
in number, were formerly included in the genus
Erica ; and some botanists still consider all the
species to belong to that genus, with the exception of those included in Calluna, while
others adopt about half the new genera. In
this uncertainty, I shall only describe two of
the doubtful genera, partly because the distinctions between them and the true heaths are
strongly marked, and partly because the species they contain are frequently met with in
British gardens and greenhouses, where they
are sometimes labelled with their old names and
sometimes with their new ones.

In the genus Erica, one of the commonest
species is the Besom Heath (*E. tetralix*), which
is found in great abundance on moorish or boggy
ground in every part of Britain. In this plant,
the corollas of the flowers appear each to consist of a single petal, forming an egg-shaped
tube (see *b* in *fig.* 47), contracted at the mouth,
but afterwards spreading into a four-cleft limb,
through which is seen projecting the style, with

its flat stigma.  The corolla is, however, really
in four petals, which, though they adhere together, may be easily separated with a pin.  The
stamens are concealed by the corolla, but the
manner in which they grow is shown at *c*; and
*d* is a single stamen, showing the spurs or awns
at the base of the anther, the position of which
is one of the characteristics of the genus Erica
in its present restricted form; *e* is a capsule
with the style and stigma attached; and *a* is a
leaf showing its revolute or curled back margin.
The leaves of this species are in whorls, four
leaves in each whorl, and they are ciliated, that
is, bordered with a fringe of fine hairs.

All the true Heaths bear more or less resemblance to this plant.  In some, the corollas are
bell-shaped, spreading out at the tip into five
teeth, which inclose the stamens, as shown in
*fig.* 48; and in others they are nearly globose
swelling out near the calyx, and tapering to a
point, beyond which the
stigma and anthers pro-
ject; as in the Cape Heath,
called *Erica hispida*, a
flower of which is shown
in *fig.* 49.  The leaves also
differ exceedingly, in the
number contained in each
whorl; as in some species
there are only three in a whorl,

FIG. 48.—
BELL-SHAPED
HEATH.

FIG. 49.—CAPE
HEATH.
(*E. hispida.*)

while in others they are five or six. The
general features of all the Heaths are,
however, the same—viz., there are eight stamens, which are generally inclosed in the
corolla, though they sometimes project beyond
it, as shown in *fig.* 49, and the anthers of
which are two-cleft, and awned or crested
at the base, while the filaments are hair-like;
one style, which always projects beyond the
corolla, and has a flattened stigma; a four-parted
calyx and corolla which is tubular, with a
four-parted limb. There are nearly two hundred
species of this genus, some of which are natives
of Europe, and others of the Cape of Good Hope.

The moor Heaths (*Gypsocallis*) were separated
from the genus Erica, by Mr. Salisbury, principally on account of the corolla being campanulate, or shortly tubular, with a dilated mouth;
and the stamens projecting beyond the corolla.
The filaments are also generally flat; the
anthers are without awns, and distinctly in two
parts; and the stigma is simple, and scarcely
to be distinguished from the style. The common Cornish Heath (*G. vagans*), and the
Mediterranean Heath (*G. Mediterranea*), are
examples of this genus, which appears strongly
marked, though, as I before mentioned, some
botanists do not adopt it.

Callista is a genus established by the late

Professor Don, which appears very distinct,
though it also has not been generally adopted.
It includes all those beautiful Cape Heaths
which have a shining, glutinous, ventricose, or
cylindrical corolla with a
spreading limb (see *a* in *fig.* 50),
and a capitate stigma (*b*). *C.*
*bucciniflora* and *C. ventricosa*,
are examples of this genus.

The Ling or Heather, which
Linnæus called *Erica vulgaris*,
is now generally placed by all
botanists in a separate genus
called Calluna, which was es-

FIG. 50.—CALLISTA
BUCCINIFLORA.

tablished by Mr. Salisbury. The calyx of this
plant is membranous, and coloured so as to
resemble a corolla, and it is furnished
with four bracts at the base, which
resemble a calyx. The true corolla
is bell-shaped, and shorter than the
calyx. The stamens are inclosed,
and the anthers are of the very
singular form shown in *fig.* 51.

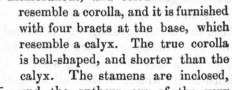

FIG. 51.—
STAMEN OF THE
LING.

The stigma is capitate, and the flowers are
disposed in what is called a racemose spike.
The leaves are trigonal; they are very short,
and they are laid over each other like scales
in four rows. The Ling is the only species in the
genus.

SUB-TRIBE II.——ANDROMEDEÆ.

THE plants in this sub-tribe differ decidedly
from those of the preceding division, in having
ten stamens, while all the genera of heaths
have only eight. The calyx is also five-cleft
instead of four; and the corolla, which falls
before the seeds are ripe, has a five-lobed
limb. The sub-tribe is divided into twenty
genera, more than half of which are perfectly
distinct.

The genus Andromeda is distinguished by its
globose corolla which has a five-lobed limb;
and its stamens which have their filaments
bearded, and their anthers short and two-
awned. *Fig.* 52 shows a stamen of
the wild rosemary (*Andromeda poli-*
*folia*) with its bearded filament (*a*),
and its two-awned anther with its
pore-like openings (*b*). The cells
of the capsule open in the middle,
down the back, to discharge the seeds. Professor
Don has divided the genus Andromeda into six
genera; some of which contain only one or two
species. Thus only *Andromeda polifolia* and *A.*
*rosmarinifolia* are left in the genus Andromeda;
Cassandra contains only *A. calyculata,* and *A.*
*angustifolia;* and Zenobia, only the beautiful

FIG. 52.—
STAMEN OF
ANDROMEDA.

I 2

*Andromeda speciosa*.    In Cassandra the anthers
are long and mutic (see *a* in *fig.* 53),
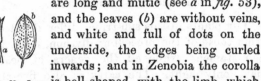
and the leaves (*b*) are without veins,
and white and full of dots on the
underside, the edges being curled
inwards; and in Zenobia the corolla
FIG. 53.—LEAF is bell-shaped, with the limb, which
AND ANTHER  is in five lobes, curling back (see *a*
OF CASSANDRA. *fig.* 54).    The stamens have the
laments (*b*) curiously dilated at the base;

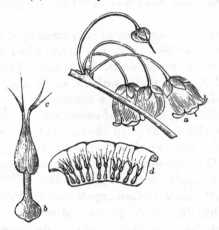

FIG. 54.—FLOWERS, COROLLA, AND STAMEN OF ZENOBIA.

and the point of each cell of the anther is cut
into two erect awns (*c*).    The manner in which
the stamens are arranged inside the corolla is

shown at (*d*).  The cells of the capsule, when
ripe, open down the centre, and the seeds
which are angular, are attached to a five-lobed
placenta.

Lyonia is a genus established by the American
botanist Nuttall, because the plants it contains
have the margins of the valves of their capsules
closed by five other narrow external valves.  The
plants are natives of North America, and their
flowers are generally small.  *Lyonia Mariana* may
serve as an example of this genus, which is generally adopted by botanists.

It would be useless to enter into details of the
other genera formed out of Andromeda, as they
are not generally adopted ; but, perhaps, it may
be worth mentioning, that the well-known *Andromeda floribunda* is placed by Professor Don in a
new genus which he calls Leucothoe.

St. Dabeoc's Heath, or Irish Whorts, a little
heath-like shrub, common in Ireland, is one
of those plants which have puzzled botanists
exceedingly.  It has been called successively
*Erica, Andromeda,* and *Menziesia, Dabœcia ;* then
*Erica Hibernica,* next *Menziesia polifolia,* then
*Vaccinium Cantabricum* and lastly *Dabœcia polifolia.*  It is probable, however, that it may
even yet be doomed to undergo other changes ;
as, from the construction of its anthers, which
are linear, and arrow-shaped at the base, and

which open lengthways, instead of by pores, it does not appear even to belong to the Ericaceæ.

The other genera in this sub-tribe are quite distinct from each other, and contain several well-known plants. The most popular of these genera are Arbutus, Arctostaphylos, Gaultheria, and Clethra.

The Strawberry tree (*Arbutus Unedo*) has little bell-shaped flowers, contracted at the mouth, and with a curling-back limb, which are easily recognised as belonging to the Ericaceæ. They have ten stamens, the filaments of which are hairy at the base (see *a* in *fig.* 55) and inserted in the disk; which in this genus is large, and rises up round the ovary (see *b*). The calyx is permanent, and five-cleft; and

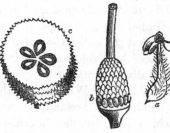

Fig. 55.—Fruit &c. of Arbutus Unedo.

the flowers are produced in panicles, and each is furnished with a bract. The fruit, which retains the calyx when ripe, is a granular berry, covered with tubercles on the outside; and it has five cells (*c*) containing the seeds. There are numerous varieties of this species common in British gardens, besides a very beautiful hybrid

between it and *A. Andrachne.* The latter
species is a native of Greece, and rather more
tender than the common kind; and it is very
conspicuous in shrubberies from its red stems
and loose bark.

The Bearberry (*Arctostaphylos Uva-Ursi*) was
formerly considered to belong to the genus
Arbutus, but it differs in the filaments of the
stamens being smooth and dilated at the base,
and the awns affixed to the middle of the
anthers. The berry is without tubercles, and
the cells are often only one-seeded.

There are two species of Gaultheria common
in British gardens: viz.—*G. procumbens* and
*G. Shallon :* both of which have flowers resembling those of the Arbutus and furnished with
bracts; but in the former species the flowers are
solitary and produced from the axils of the
leaves, and in the latter they are in racemes, of
the kind called secund, that is with the flowers
growing all on one side. The berries of both
kinds are eatable, and those of *G. procumbens*
are called Partridge berries in America, and
the leaves Mountain tea. Both species have
ten stamens, the anthers of which are two-
cleft, each cell being furnished with two horns,
as in *Zenobia speciosa* (see *fig.* 54, in page
116). The fruit is five-celled and the seeds are
numerous.

The genus Clethra differs considerably from the preceding genera, as the limb of the corolla is so large and so deeply cleft, as to make the flower appear to have five petals (see *a* in *fig.* 56). There are ten stamens, with broad arrow-shaped anthers (*b*), and a three-cleft stigma, (*c*). The capsule is dry, with three many-seeded cells. In *C. alnifolia*, a native of North America, (of which *fig.* 56 represents a magnified flower,) the flowers are erect, and produced in a

FIG. 56.—FLOWER OF CLETHRA ALNIFOLIA.

spicate raceme; but in *C. arborea*, a native of Madeira, the racemes are panicled, and the flowers drooping and somewhat bell-shaped. Both species are very ornamental.

---

### TRIBE II.—RHODOREÆ.

THE plants included in this tribe are all considered to bear more or less resemblance to the Rhododendron, though in some of them the family likeness is not very strong; and the genera I shall describe to illustrate it are

Rhododendron, Azalea, and Rhodora (the last two being by some botanists included in Rhododendron); Kalmia, Menziesia, and Ledum.

The species of the genus Rhododendron are easily distinguished by their flower buds, which are disposed in the form of a strobile, or pine-cone, each bud having its accompanying bract, which the flower retains after its expansion, as shown in *fig.* 57 at *a*, in a flower of *R. maximum*.

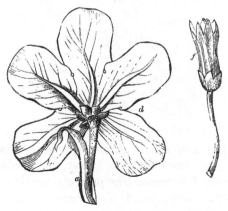

FIG. 57.—BACK VIEW OF A FLOWER OF RHODODENDRON MAXIMUM, AND SEED-POD.

There are five or ten stamens of unequal length, the larger ones curling upwards (as shown at *b* in *fig.* 58), as does the style (*c*), which has a simple stigma. The flowers have a very small calyx, (*d* in *fig.* 57,) and a campanulate corolla which is deeply five-cleft, the upper segment (*e* in *fig.* 58) being somewhat larger than the rest, and spotted in the inside. The capsule is five-celled and five-valved, as shown in *fig.* 57 *f.*

FIG. 58.—FLOWER OF RHODODENDRON PONTICUM.

The leaves of nearly all the species are ever-green; and the flowers are showy, and produced in terminal corymbs. The principal species may be thus distinguished from each other; *R. maximum* has drooping leaves, covered with brown or white down on the under surface, and a dense corymb of flowers, the segments of the corollas of which are roundish, and the bracts leafy. In *R. ponticum*, on the contrary, the corymbs of flowers are looser, the segments

more pointed, and the bracts more scale-like ;
and the leaves are smooth on both surfaces.
The seed-pods also differ: in those of *R. maximum*
and the other American species, the valves are
smooth as shown at *f* in *fig.* 57 ; and in those
of *R. ponticum*, the valves are somewhat crin-
kled as shown in *fig.* 59.　This
species, and all its hybrids and
varieties, are more tender than *R.*
*maximum*, *R. catawbiense*, and all
the other American kinds and their
offspring.　*R. catawbiense* has the
flower of a darker colour on the
outside of the corolla than within,
and the upper segment is very faintly
dotted.　It hybridises freely with *R.*
*arboreum*, which *R. maximum* does not, and the
hybrids thus produced are hardier than those
raised from *R. ponticum*, though the latter are
by far the most numerous.

FIG. 59.—SEED-
POD OF
RHODODENDRON
PONTICUM.

　　Most of the species have purple or whitish
flowers, but some, such as *R. chrysanthemum*,
and *R. anthopogon*, have yellow flowers ;　*R.*
*ferrugineum* and *R. hirsutum*, have bright pink
or rose-coloured flowers ;　and those of *R.*
*arboreum* the Nepaul tree Rhododendron, are of
a rich scarlet.　The commonest small kinds
are *R. ferrugineum* and *R. hirsutum*, both dwarf
shrubs and natives of the north of Europe,

with funnel-shaped corollas, and leaves dotted
on the under surface. They are so much alike
as scarcely to be distinguished at first sight,
but on examination the leaves of *R. ferrugineum*
will be found to have brown dots, and to be
plain on the margin; while those of *R. hirsutum* have white dots and are fringed with fine
hairs.

Of all the species of the genus, those which
differ most widely from the others are the
Indian kinds. Of these *R. arboreum* has a ten-
celled capsule, and the segments of the corolla
two-lobed with waved margins. The leaves are
long and silvery beneath; and the capsules, the
peduncles, and the calyxes, are all woolly. In
*R. campanulatum*, a splendid species with very
large flowers, the capsule is six-celled, the leaves
are somewhat cordate at the base, and the
bracts are fringed; and in *R. anthopogon* the
corolla has a cylindrical tube, woolly inside, and
a small but spreading limb, cut into five lobes.
There are eight stamens, and the capsule is five-
celled.

*R. Camtschaticum, R. Chamæcistus,* and *R.
dauricum* differ from the preceding species in
having their corollas rotate, that is, wheel-
shaped. The last of these kinds is a favourite greenhouse shrub, from its flowering
under shelter in winter. In the open ground

it flowers in March. The species has rose-coloured flowers which appear before the leaves; and leaves which turn red in autumn before they fall. The roots are knobbed and fibrous; and the stems are twisted and knobbed in a wild state. There is a variety *R. d. atrovirens* which has purple flowers, and evergreen leaves, and which is hardier than the species.

The genus Azalea may be divided into three kinds, viz., *A. indica* and its allied species; *A. pontica* and its varieties and hybrids; and the American Azaleas. These divisions are easily distinguished by their flowers. Those of the Indian or Chinese Azaleas have all large showy flowers, on short downy footstalks, and they are produced in small clusters of only two or three flowers each, at the extremity of the shoots. The corollas are bell-shaped and deeply cut, nearly to the base, into broad spreading segments. The stamens are ten in number, shorter than the corolla, and of unequal length. The leaves are evergreen, and they are numerous, thickly set and downy. These Azaleas are all very handsome, but the white Indian Azalea (*A. indica alba*, or *A. ledifolia*) is particularly so, and very fragrant. The species belonging to this division are mostly natives of China, and require either a greenhouse or some slight protection during winter in England.

The yellow Azalea (*A. pontica* or *Rhododendron flavum*) differs from *A. indica* in being quite hardy; in the flowers being produced in umbels of from eight to twelve, at the ends of the branches, before the leaves; and in the corollas being funnel-shaped instead of campanulate. The tube of the funnel is, however, shorter than the limb, the segments of which are broad and spreading, the upper three being larger and of a darker yellow than the two below. There are usually five stamens, projecting a little beyond the corolla, and curving upwards; the style also curves upwards, and it is crowned by the stigma, which forms a round green head.

The calyx is very small, and both it and the corolla feel clammy to the touch. The flowers are fragrant. The leaves are deciduous, and they are ovate, slightly hairy, and terminate in a mucro or stiff point. There are many varieties of this species, and many hybrids between it and the American kinds, all of which are quite hardy in British gardens.

The principal American Azaleas are *A. nudiflora, A viscosa, A. nitida,* and *A. speciosa,* all of which have the corollas of their flowers funnel-shaped. Of these *A. nudiflora* is easily known by its stamens, which project a long way beyond the corolla, and by the tube of the corolla being longer than the limb. The plant is deciduous; and the flowers, which are produced in large terminal clusters, and which are not clammy, appear before the leaves. The common English name for this plant in some parts of the country is the American Honeysuckle, and the flowers are of various shades of red, pink, white, and purple. *A. calendulacea,* which some botanists make a variety of this species, has much larger flowers, and the leaves pubescent on both surfaces, whereas, in *A. nudiflora* the leaves are nearly smooth and green, with only a slight fringe of hairs round the margin. There are numerous varieties of *A. calendulacea,* the flowers of which are always either yellow, red, orange, or copper-coloured, and it is supposed to be the parent of the beautiful Ghent Azaleas. *A. viscosa* has the tube of the corolla equal in length to the limb, and rather short stamens; the flowers of this species are clammy. *A. hispidum,* which is generally considered a variety of *A. viscosa,* is still more clammy, and the tube of the corolla is wider and shorter; other probable varieties are *A. nitida,* which has shining leaves, and *A. glauca,* which has glaucous ones, as in both kinds the flowers are very clammy. *A. speciosa* has large flowers and leaves tapering at both ends. All the species of Azalea have five stamens, but some of the varieties have ten.

*Rhodora canadensis* is a little American shrub

with pink flowers, which appear before the leaves, and the corolla of which is bilabiate, the upper lip being the broadest, and cut into two or three teeth, and the lower only once cut. There are ten stamens, and the capsule is five-celled and five-valved. The leaves are deciduous, and slightly pubescent beneath ; and the flowers are produced in small terminal clusters. This plant, as well as all the Azaleas above described, are now included by some botanists in the genus Rhododendron.

The genus Kalmia also belongs to this tribe. The flowers of this well known shrub are very curiously constructed. The corolla is salver-shaped, that is, nearly flat, and on the under side of the limb are ten protuberances, producing as many hollows on the upper side, in which lie half-buried the ten stamens. This singular construction gives the corolla that wrinkled appearance which has procured for the plant its American name of Calico flower ; while, from the shape of the leaves, it is also frequently called the Mountain laurel; it is also called Sheep laurel from its being considered poisonous to those animals when they feed on it. There are several species, which differ from each other principally in the shape of their leaves and the size of their flowers.

*Menziesia* is a genus containing only three

species, of which *M. pilosa* (*fig.* 60) may be taken as an example. The flowers are small and bell-shaped, and the anthers (*a*) are without any awns or bristles; there are eight stamens,

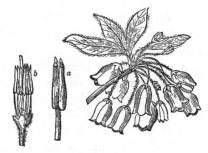

FIG. 60.—FLOWERS, ANTHERS, AND PISTIL OF MENZIESIA.

and the curious manner in which they are crowded round the style is shown at *b*. The capsule is four-celled.

*Loiseleuria*, or *Azalea procumbens*, is a small plant, having the appearance of thyme, which is the only species left in the genus Azalea by those botanists who include the true Azaleas in the genus Rhododendron.

*Ledum* is the last genus belonging to this tribe that I shall attempt to describe. *Ledum palustre*, or wild Rosemary, the best-known species, has a corolla in five regular petals, and ten stamens which project beyond it; but *L. latifolium*, the Labrador Tea, has only five stamens, which are

K

not longer than the petals. *L. buxifolium*, a little thyme-like shrub, is now called *Leiophyllum thymifolium*. All the species have white flowers.

---

### TRIBE III.—VACCINIEÆ.

THE plants comprised in this tribe, which is considered a separate order by many botanists, all agree with the genus Vaccinium in having the ovary entirely surrounded by the calyx, which forms a fleshy berry-like fruit when ripe, and in the seeds being scaly. *Vaccinium Myrtillus*, the common Bilberry or Blaeberry, is a familiar example of the genus; and

FIG. 61.—COMMON BILBERRY (*Vaccinium tenellum*).

*fig.* 61 shows the shape of the flowers at *a*, the manner in which the ovary is enveloped in the calyx at *b*, and the curious shape of the anthers in the magnified representation of them at *c*. The berry is five-celled and many-seeded; and there are eight or ten stamens. Both the anthers and the flower vary in the different species, but the calyx and the manner in which it surrounds the ovary are nearly the same in all, as may be seen in *fig.* 62, which represents a specimen of *V. tenellum*, the Pennsylvanian Whortle-berry.   In this figure *a* is the flower, *b* the anther, and *c* the ovary surrounded by the calyx.

FIG. 62.—AMERICAN WHORTLE-BERRY (*Vaccinium pennsylvanicum* .

There are many species, among which may be mentioned the American Bluets (*V. angustifolium*); Deerberries (*V. stamineum*); Bluetangles (*V. frondosum*); the Hungarian Whortle-berry (*V. Arctostaphylos*); and the Cow-berry, or common British Whortle-berry (*V. Vitis-Idæa*).

The Cranberry (*Oxycoccus palustris*) differs from the genus Vaccinium in the shape of its flowers (see *fig.* 63), and in its anthers being without spurs; there are eight stamens, the filaments of which are connivent, that is, growing close together. The American Cranberry (*O. macrocarpus*) differs from the European kind, principally in having larger fruit.

FIG. 63.—CRANBERRY.

K 2

### TRIBE IV.—PYROLEÆ.

THIS tribe is also considered as a separate
order by many botanists; but the principal distinction is the long arillus or skin which enfolds
the seeds and gives them the appearance of being
winged. The most remarkable genera are *Pyrola*,
the Winter Green, of which there are several
species common in moist woods in the north of
England and Scotland ; and *Monotropa*, or
Bird's-nest, parasitic plants which grow on
the roots of pine and beech-trees, but are by
no means common in England. The species of
Pyrola are pretty little evergreen plants, with
white flowers, the corollas consisting of five
distinct petals, and which have ten stamens, the
anthers of which are two-celled, each opening
by a pore ; the style is single, ending in a capitate
stigma cut into five lobes; and the capsule is
five-celled. The yellow Bird's-nest, (*Monotropa
Hypopitys*) has a coloured stem, with drooping
flowers, and numerous scales instead of leaves,
of which it is destitute. The flowers have a
coloured calyx cut into four or five segments,
and the corolla is in four or five petals. There
is an American species with white flowers.

## CHAPTER VIII.

THE ORDER OLEACEÆ, OR JASMINEÆ : ILLUSTRATED BY THE COMMON WHITE JASMINE; THE YELLOW JASMINE; THE PRIVET; THE PHILLYREA; THE OLIVE; THE FRINGE-TREE (*Chionanthus Virginica*); THE LILAC; THE COMMON ASH; AND THE MANNA OR FLOWERING ASH.

THIS order was established by Jussieu, who divided it into two tribes—Jasmineæ and Oleineæ, which are now very generally considered as distinct orders. I have, however, thought it best to keep them together, as I wish to make as few divisions as possible, to avoid burthening the memory of my readers. All the genera in both tribes agree in their flowers having only two stamens, an ovary with two cells, and two seeds in each cell; and anthers with two cells, which open with a long slit lengthways.

The species of the Ash have no corolla; but in all the genera where there is one, the filaments of the stamens, which are very short, are inserted in it; and it is generally funnel-shaped—as, for example, the corolla of the Jasmine. Though the ovary is two-celled, and the cells two-seeded, each flower very often only produces one perfect seed. The leaves are generally pinnate.

### TRIBE I.—JASMINEÆ.

THE genus Jasminum is the only one in this
tribe which contains plants common in British
gardens; and of all the species contained in it,
the common white Jasmine (*J. officinale*) is
perhaps the best known. The flowers are produced in terminal clusters of four or six. The
calyx is tubular, with the limb cut into numerous
narrow segments; (see *a* in *fig.* 64;) and the

FIG. 64.—FLOWER AND LEAF OF THE JASMINE.

corolla is funnel-shaped, with a spreading limb
(*b*) divided into four or five pointed segments,
which are folded over each other, and somewhat
twisted in the bud. The two stamens and the
style and stigma are enclosed in the corolla;
and the fruit is a berry divided into two cells,
with one seed in each. There is no albumen

in the seeds.  The leaves (*c*) are impari-pinnate,
with the single terminating leaflet larger than
the others; and the petioles are articulated.
The common yellow Jasmine (*J. fruticans*) has
flowers in terminal clusters of three each, and its
leaves are either ternate, that is, with three
leaflets, or simple.  The branches are angular,
and the leaves quite smooth.     The Nepaul
yellow Jasmine, (*J. revolutum*) has pinnate
leaves of five or seven leaflets, which are smooth
and shining.  The flowers are large and produced in compound corymbs.  They are a bright
yellow, and very fragrant.  The segments of
the corolla are obtuse, and the stigma club-
shaped.  There are above seventy species of
Jasmine, more than twenty of which have been
introduced into Britain; but they may be all
easily recognised by their flowers, which bear a
strong family likeness to each other, and by
the petioles of their leaves, which are always
articulated or jointed, that is, they will break
off the stem without tearing the bark.   In
other respects the leaves vary exceedingly in
this genus, some being simple and others compound; and some being opposite, as in the
common Jasmine, and others alternate, as in
*J. revolutum.*

## TRIBE II.—OLEINEÆ.

THIS tribe contains numerous genera, among
which the most common are the Privet (*Ligustrum*), Phillyrea, the Olive (*Olea*), the Fringe-
tree (*Chionanthus*), the Lilac (*Syringa*), the Ash
(*Fraxinus*), and the Flowering or Manna Ash
(*Ornus*). All these genera agree in their general
character with Jasminum, except as regards
their seeds, which abound in albumen.

In the common Privet (*Ligustrum vulgare*),
the flowers, which are produced in terminal
compound racemes, have a very short calyx
(see *a* in *fig.* 65), with a funnel-
shaped corolla, having a wide tube
in proportion to the limb (*b*), which
is very short and divided into four
segments.    The anthers of the
stamens and the stigma are seen in
the throat of the corolla.    The
berry is drupe-like, and generally
contains two one-seeded nuts.  The
leaves are simple and opposite.    There are many
species of Privet, but the handsomest is *L.
lucidum*, the leaves of which are broad and
shining, and the panicles of flowers spreading.
This tree yields a kind of waxy matter from its
leaves and branches when boiled, which is said
to be used by the Chinese for candles.

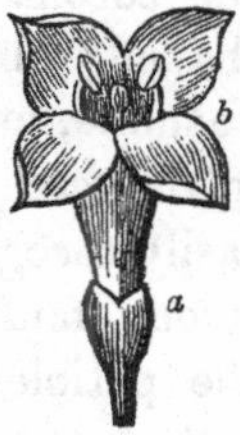

FIG. 65.—FLOWER
OF THE PRIVET.

The Phillyrea is a handsome evergreen shrub, very useful in shrubberies, from its forming a close compact bush of a deep green, which makes a good background to Tree Roses, Almond-trees, *Magnolia conspicua*, or any other flowering plant that would appear naked if its flowers were not relieved by a back-ground of green. The flowers of the Phillyrea are small and of a greenish white. The fruit is a drupe, containing a two-celled stone or nut, but with seldom more than one perfect seed.

The Olive (*Olea sativa*) has small white flowers, resembling those of the Privet, and a fleshy drupe like a Sloe, with a one or two celled stone or nut. The oil is contained in the fleshy part of the fruit, and the best oil is that which is obtained by crushing the pulp of the fruit without breaking the stone or nut.

The Fringe-tree (*Chionanthus virginica*) differs from the preceding genera in the length of the segments of the limb of its corolla, which is cut into long slender shreds like fringe. In all other respects except that the pulp of the fruit does not contain oil, this genus is closely allied to the Olive.

The common Lilac (*Syringa vulgaris*) has its flowers disposed in a kind of panicled raceme called a thyrsus. The calyx is very small, and obscurely four-toothed (see *a* in *fig.* 66), and the corolla (*b*) is funnel-shaped, with a four-parted limb ; the stigma is two-cleft, and both the style and stamens are enclosed in the tube

FIG. 66.—FLOWER AND SEED-PODS OF THE LILAC.

of the corolla. The fruit is a dry two-celled and two-seeded capsule, which opens with two valves, as shown at *c*, each valve having a narrow dissepiment down the middle : the shape of the seed is shown at *d*. The leaves are simple, opposite, and entire ; and the branches are filled with pith, which may easily be taken out and the branch left hollow like a pipe ; and hence the generic name of Syringa, from *Syrinx* a pipe.

The Ash (*Fraxinus excelsior*) differs so much from the other genera as to seem scarcely to belong to the same order. The flowers are without any petals, and frequently without any calyx ; and some of them, which are called the female flowers, have no stamens, while others, which are called the males, have no pistil. Some of them, however, have both stamens and pistil.

The fruit is what is called a samara or key ; that is, it is furnished with a membrane-like wing so as to resemble a dry leaf. It is two-celled, but very frequently only one-seeded. The shape of the keys, and the manner in which they grow, is shown at *a* in *fig.* 67; and the leaves, at *b*. The leaves are opposite and generally pin-nate, with five or six pairs of leaflets; but there is one species with simple leaves (*Fr. simplicifolia*). The Weeping Ash

FIG. 67.—AMERICAN ASH (*Fraxinus americana*).

is only an accidental variety of the common kind. The leaves of the Ash come out late and fall early ; but the tree may easily be recognised when quite bare by the greyness of its bark and its black buds. It will grow in any soil ; but it is injurious to arable land, from its roots spreading widely near the surface.

The Manna, or flowering Ash, (*Ornus euro-pæus*), differs widely from the common Ash in its flowers, which are white, with a corolla divided into four long narrow segments. The two stamens have long filaments, with a small pistil (*c*), the stigma of which is notched. The

flowers are produced in great profusion in loose
panicles, and they are very ornamental, the
samaras and leaves closely resembling those
of the common ash.　There are several speci es
of this genus, which were all formerly included
in the genus Fraxinus.　The Manna is the
sap of the tree, and it is procured by wounding
the bark.

# CHAPTER IX.

THE ORDER SOLANACEÆ: ILLUSTRATED BY THE BITTER-SWEET;
GARDEN NIGHTSHADE ; POTATO ; EGG-PLANT ; TOMATO ; CAPSI-
CUM ; WINTER CHERRY ; CAPE GOOSEBERRY ; THE DEADLY
NIGHTSHADE ; LYCIUM, OR DUKE OF ARGYLE'S TEA-TREE ;
CESTRUM ; VESTIA ; TOBACCO ; PETUNIA ; NIEREMBERGIA ;
SALPIGLOSSIS ; SCHIZANTHUS ; HENBANE ; DATURA ; BRUG-
MANSIA ; SOLANDRA ; VERBASCUM ; CELSIA ; NOLANA ; ETC.

THIS large Order is one of those which
appear to have been most troublesome to botanists, as scarcely any two agree as to the plants
to be comprised in it. I have, however, taken
it in its most comprehensive sense, as far as
popular plants are concerned ; on the same
principles as those by which I have been
guided throughout ; viz. that it is easier for
a beginner to remember a few divisions than
a great many ; and that when a student has
once learnt what plants are nearly allied to
each other, and the general features that connect them, it will be comparatively easy to learn
the minor distinctions between them.

Taking these principles as my guide, I have
given the Order Solanaceæ as it was formed by
Jussieu, adding those plants to it which evidently
belong to the several sections, but which have

been discovered since the time of that great
naturalist; and I have divided the Order into
four tribes, viz. Solanaceæ, Nicotianeæ, Verbascineæ, and Nolaneæ. All these plants agree in
having the stamens, which are generally five,
inserted in the corolla, the calyx and corolla
inclosing the ovary, and the calyx remaining on
the ripe fruit.

---

### TRIBE I.—SOLANACEÆ.

The plants included in this tribe are easily
recognised by their flowers, which bear a considerable resemblance to each other, and by
their berry-like fruit, which has always a persistent calyx. The corolla is also always folded
in the bud; and the folds, like those of a country
woman's clean apron, are often so deeply im-
pressed as to be visible in the newly opened
flowers. The genera included in this tribe
differ widely in their qualities.

The genus Solanum is easily recognised by a
botanist through all its numerous species by its
anthers, which open by two pores like those of the
Ericaceæ, and which differ in this respect, from
the anthers of all the other plants contained in
the Order, all of which open by a long slit down
each cell. The flowers of all the species of Solanum
are of the kind called rotate, or wheel-shaped;

but they are generally cut into five distinct
segments: which are sometimes turned back, as
in the flower of the Bitter-sweet (S. *Dulcamara*),
as shown in *fig.* 68 *a*; and sometimes nearly flat,
as in the flower of
the common garden
Nightshade (*S. ni-
grum*). The berries
of the Bitter-sweet
(*b*) are red, and they
have a very pretty
effect in hedges and

FIG. 68.—BITTER-SWEET (*Solanum
Dulcamara*).

wild coppices, where they are produced in great
abundance during the latter part of summer
and autumn ; and those of the Garden Nightshade are black. Both these plants are poisonous ;
but this is by no means the case with all the
species of the genus, as the tubers of the potato
(*S. tuberosum*) are, as is well known, wholesome food, and the fruit or apple is not decidedly
poisonous ; while the Aubergine, or Egg-plant
(*S. Melongena*), which is another species, has a
fruit which is large, smooth and shining, and
which when boiled or stewed is good to eat.
The segments of the corolla of this species are
often so deeply notched as to appear to be six
or nine, instead of five.

There are many ornamental species of Solanum, many of which have woolly, and some

prickly leaves; but the flowers have all such a likeness to each other, as seldom to require to be botanically examined to be recognised.

The Tomato or Love-apple, (*Lycopersicum esculentum,*) has flowers which bear a great resemblance to those of some of the species of Solanum, but the anthers open longitudinally and are connected by a membrane into a kind of cylinder. The seeds also are hairy; and the berry is wrinkled, and not of so firm a texture as in Solanum. The flowers of this plant are frequently united, so as to appear to have double or treble the usual number of stamens, and two or three styles; and when this is the case, the fruit appears deformed from two or three of the ovaries having grown together. The fruit is very good to eat, and wholesome either boiled or stewed, or as sauce. There are several species, all of which were formerly included in the genus Solanum.

The plants belonging to the genus Capsicum have flowers which are very much like those of the Tomato, and which have similar anthers; but the fruit differs in being a dry, inflated, hollow berry, inclosing numerous seeds, and in both the seeds and their cover having a fiery biting heat to the taste. There are several species with fruit of greater or less size, and different colours; generally red or

yellow, but sometimes white or green. The
best Cayenne pepper is made from the pods of
*C. frutescens*, dried in an oven and then reduced
to powder. The annual species (*C. annuum*) has
many varieties; one of which produces the small
pods called by the market-gardeners Chilies,
and which are eaten fresh by dyspeptic patients,
to assist digestion.

The Winter Cherry (*Physalis Alkekengi*) has
the same kind of flower as the other genera of
this tribe. The corolla is rotate, and obscurely
five-lobed; and the stamens, which are connivent, (that is, lying close together), have very
large anthers. When the corolla falls, the calyx
becomes inflated, and expands to a large size,
completely enclosing the little berry-like fruit in
the centre. A very beautiful preparation may be
made by soaking this calyx in water till it becomes
completely macerated ; that is, till all the pulp
is decayed and only the fibrous part left. The inflated calyx then appears like a beautiful network
covering, with the bright red berry in the centre.
To macerate the calyx properly, it should be
left in the same water without changing, for
about six weeks. The Cape Gooseberry (*P.
peruviana*) is another species of the genus
Physalis ; but instead of being a native of
Europe, it is from Peru; and its flowers, instead
of being white, are yellow, with a dark red spot

at the base of each lobe of the corolla: the
berry also is yellow. This species is called
Cape Gooseberry, because it is cultivated as a
fruit at the Cape of Good Hope.

The Deadly Nightshade (*Atropa Belladonna*)

differs widely from
all the preceding
genera in having
a bell-shaped co-
rolla, (see *a* in *fig.*
69,) and in the an-
thers (*b*) not lying
close together. It
has, however, a per-
manent calyx and
a two-celled berry,
like the rest.

The Barbary
Box-thorn, or Duke
of Argyle's Tea-

Fig. 69.—DEADLY NIGHTSHADE
(*Atropa Belladonna*).

tree, (*Lycium barbarum*) has a somewhat rotate
corolla, with a five-cleft limb, with the stamens
inserted between the segments in the same manner as shown in the flower of the *Atropa Belladonna*, represented cut open at *b* in *fig.* 69. The
filaments are hairy at the base, and the anthers
are near together, but do not form a cone as in
Solanum. The berry is two-celled, and the
calyx remains on when it is ripe, as in all the

other genera of this order. There are several
species of Lycium, which are all known by the
English name of Box-thorn; but *L. barbarum* is
also called the Duke of Argyle's Tea-tree, from
a story told of this plant being sent to a
Duke of Argyle early in the last century, instead
of the true Tea-tree. The story, however,
is very doubtful; and the more so, as in France,
the dwarf Chinese Elm is called Thé de l'Abbé
Gallois, as it is said, from a similar cause.

*Cestrum Parqui* is a very handsome half-hardy
shrub, which may be placed in this division
from its berry-like fruit. It has a funnel-shaped
corolla, with a five-lobed limb, enclosing its five
stamens. The flowers are disposed in an upright
raceme; they are yellow, and very fragrant.
The berries are of a very dark blue, and almost
black when ripe. *Vestia* is another genus very
nearly allied to Cestrum, but the stamens project
beyond the mouth of the corolla instead of
being enclosed within it; and the flowers, which
are produced singly, have a very disagreeable
smell.

---

### TRIBE II.—NICOTIANEÆ.

THE plants included in this tribe agree with
those of the preceding division, in having the
corolla generally folded in large plaits in the

bud; but they are distinguished by having all
capsular fruit: that is, in all the plants
belonging to this tribe, the seed-vessel is dry
and hard when ripe, and not soft and pulpy
like a berry. The species have nearly all
funnel-shaped flowers, with a long tube and a
spreading limb; the tube is generally very
long in proportion to the limb, and it is often
inflated, so as to appear much wider in the
upper part than near the calyx.

The Virginian Tobacco (*Nicotiana tabacum*) is
an example of an inflated tube to the corolla

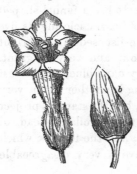

(see *fig.* 70, *a*); the limb
is small and divided
into five pointed seg-
ments; and the cap-
sule (*b*), which opens
at the point into four
valves when ripe, con-
tains numerous seeds.
The whole plant is
covered with a clammy
down, particularly the
leaves, which are large

FIG. 70.—VIRGINIAN TOBACCO.

and flabby, and which have their footstalks
dilated at the base, so as partly to enfold the
stem. There are many species of Nicotiana,
some of which are very ornamental. It is the
dried leaves that are used as tobacco, or ground
into snuff.

The Petunias are so well known, that I need
say very little of the general form of their
flowers, except to point out the connexion between them and the Tobacco. The corolla is
salver-shaped, with a cylindrical tube, wider at
the top than at the base, and a five-lobed limb.
There are five stamens of unequal length, which
are hidden in the tube of the corolla. The
stigma has a broad head which is slightly two-
lobed ; and the calyx remains on the ripe capsule, which is two-celled, and opens in the upper
part with two valves. The seeds are numerous
and very small, and the leaves are pubescent
and slightly clammy. If my readers will take
the trouble to compare the Petunia and the
Tobacco, they will be surprised to find how much
the flowers are botanically alike. The differences
are, that the calyx is more leaf-like in the Petunia
than in the Tobacco ; and the corolla of the
Petunia is somewhat oblique, that is, two of the
segments are smaller than the others ; the filaments, also, are thickened at the base. It will
appear extraordinary to every one acquainted
with the flowers of the purple and the white
Petunias, to find that some botanists have placed
them in different genera. Such, however, is the
case. On cutting open the delicate little seed
of the white Petunia (*P. nyctaginiflora*), which
it must have been very difficult to do, and ex-

amining it in a very powerful microscope, the
embryo or germ of the future plant was found
to be curved like that of most of the other
Solanaceæ; whereas when the seed of the
purple Petunia (*P. phœnicea* or *violacea*) was examined in the same manner, the embryo was
discovered to be straight. This purple Petunia
has consequently puzzled botanists as much as
some of the other plants I have had occasion
to mention; and it has been called successively
*Petunia violacea, Salpiglossis integrifolia, Nierem-
bergia phœnicea*, and *Petunia phœnicea.*

*Nierembergia* is a genus of ornamental greenhouse plants, easily distinguished from the
Petunias by the great length of the tube of the
corolla, and by the equal segments of the limb.
The stamens also project beyond the flower,
being inserted in the throat of the corolla, and
the filaments grow together at the base; the
stigma, is, likewise, curiously dilated into a kind
of crescent shape, and it is folded in a very singular manner round the filaments, as if to support the anthers. The most common species of
this genus are *N. filicaulis, N. calycina*, and *N.
gracilis.*

The genus *Salpiglossis* is now confined to one
species, *S. sinuata*, so called from its notched or
scolloped leaves; all the different kinds being
now considered only varieties. The calyx in

this species is five-angled and five-cleft, and the corolla is funnel-shaped, the tube being very narrow near the base, and spreading out wider towards the mouth. The limb is five-cleft, and there are five stamens, one being much smaller than the others. The stigma is transverse, with a channel through the centre.

*Schizanthus* is another genus nearly allied to the last, but it is more difficult to give a just idea of it than of any other that I have attempted to describe. All the parts of the flower are irregular. The segments of the calyx are uneven; and the limb of the corolla is cut into a number of irregular lobes. There are only two perfect stamens, but there are two other small ones without any pollen in their anthers, and the rudiments of a fifth. The two perfect stamens are very elastic, springing upwards and discharging their pollen at the slightest touch. The capsule is two-celled, the valves opening at top; and the leaves are bi-pinnatifid.

The genera Salpiglossis and Schizanthus have been removed by Dr. Lindley from Solanaceæ, and placed by him in the allied order Scrophularinaceæ, or the Foxglove family.

The Henbane (*Hyoscyamus niger*) has the calyx ventricose at the base, and the corolla campanulately funnel-shaped; the limb is five-cleft, and one of the segments is larger than the

rest ; but the most remarkable part of this plant
is the capsule. When the corolla falls, the
capsule shrouded in the calyx presents the appearance shown at *a* in *fig.* 71 ; and as the
seeds ripen, the upper
part (as shown at *b*)
becomes detached,
and opens like a little
cap. The leaves are
sinuated and semi-
decurrent. There are
several species of
Henbane, one of
which (*H. aurea*) has
the limb of the corolla
deeply cut on only
one side.

FIG. 71.—HENBANE.

The genus *Datura* is nearly allied to *Brugmansia*, but it may be easily distinguished by
its calyx, which divides when the ovary begins
to swell, and the upper part drops off, leaving
only the lower part to enfold the capsule. The
corolla of all the species of Datura is funnel-
shaped, and the limb, in the large-flowered kinds,
often shows marks of the plaits in which it lay
when it was folded in the bud. There are five
distinct stamens, which are generally enclosed
in the mouth of the corolla. The capsule is
fleshy when young, and in most of the species it

is covered with spines. This is the case with
*D. Stramonium* (the common Thorn-apple), *D.
Tatula*, and *D. Metel*, all of which have also
their stamens enclosed ; but in *D. ceratocaulon*
the capsule is smooth and the stamens exserted,
that is, they project beyond the tube of the
corolla.

The genus Brugmansia is distinguished by its
calyx being ventricose, and only two or three
cleft ; it is also strongly ribbed. The corolla is
funnel-shaped, the tube being strongly ribbed ;
and the limb is five-lobed, the lobes being cuspidate, that is, drawn out into abrupt points.
The flowers are drooping, and in *Brugmansia
suaveolens*, formerly *Datura arborea*, they are
very fragrant. The anthers grow together.
The capsule is two-celled, smooth, and of a
golden yellow, and the seeds are each covered
with a thick corky skin. In *Solandra*, a nearly
allied genus, the calyx bursts on one side, and
the lobes of the corolla are not cuspidate, but
rounded and fringed. The stamens also project
beyond the mouth of the corolla, and the capsule is four-celled. The species of Solandra are
all stove-trees.

---

TRIBE III.—VERBASCINEÆ.

THE plants included in this division differ
from those in the preceding ones, in not having

the corolla plaited in the bud, and in having
the anthers only one-celled ; distinctions which
have been thought of sufficient importance to
induce many botanists to make this tribe a
separate order.

The British plant sometimes called the Shepherd's Club, and sometimes the common Mullein
or Flannel flower (*Verbascum Thapsus*), is a
familiar example of this genus. In this plant
the flower is rotate, or wheel-shaped, and divided
into five rather unequal lobes. The calyx is
five-cleft; and it possesses such a power of
collapsing over the ovary, that when the stem of
the plant is struck sharply with a hard substance, every open flower is forced off by the
sudden closing of its calyx. There are five
stamens, the filaments of which are bearded,
and the anthers crescent-shaped; and a capsule,
the two cells of which frequently run into one,
and which opens by two valves at the apex.
The flowers are crowded together in a thick
spike-like raceme, which bears no small resemblance to a club. This plant was formerly supposed to be efficacious in driving away evil
spirits ; and hence it was called Hag's-taper, now
corrupted to High-taper. The whole plant is
mucilaginous, and a decoction of it is often given
to cattle when they are suffering under pulmo-
nary complaints ; and hence is derived another

of its names, Cow's Lungwort. The leaves are
thick, and woolly on both sides; and they are
decurrent, that is, running down the stem, like
little wings on each side.

*Celsia* differs from Verbascum botanically in
having only four perfect stamens, two of which
are shorter than the others. The racemes are
also much more loose, from the flowers being on
rather long pedicels. Most of the species composing this genus were formerly included in
Verbascum. *Ramonda* is another genus, which
consists only of the *Verbascum Myconi* of
Linnæus.

---

### TRIBE IV.—NOLANEÆ.

THIS tribe, which is now made a distinct
order by Dr. Lindley, is principally known by
the genus Nolana; the species of which are
annual plants, natives of Chili and Peru, which
have lately been much cultivated in British
gardens. The flowers of *Nolana atriplicifolia*,
one of the commonest kinds, very much resemble
those of the common *Convolvulus tricolor*, and
the leaves are large and juicy like those of
spinach. On opening the corolla there will be
found to be five stamens, surrounding four or
five ovaries, which are crowded together on a
fleshy ring-like disk. These ovaries, when ripe,
become as many drupes, enclosing each a three

or four celled nut or bony putamen, which is
marked with three or more grooves on the outside, and has three or more little holes beneath.
All the species of Nolana have the same peculiarities in their seed-vessels, though they differ
in many other respects.  In the same tribe or
order are included two other genera, one of
which, called Grabowskia, contains only the
singular shrub formerly called *Lycium boer-
haviæfolium*, or *Ehretia halimifolia*, the nuts of
which resemble those of the Coffee.

Besides the plants contained in these four
tribes, there are several other genera which
some botanists place in Solanaceæ, and others
in Scrophularineæ ;  and among these may be
mentioned Franciscea, Browallia, and Anthocercis.   In the former of these genera the
flowers are small, the corolla is salver-shaped,
and the calyx, which is permanent, is inflated
and smooth.   In Browallia, the calyx is strongly
ten-ribbed, and the corolla has an oblique limb ;
and in both genera there are only four stamens,
two of which are longer than the others.   In
Anthocercis there are four perfect stamens and
the rudiments of a fifth.   The corolla is not
folded in the bud, but has a regular, star-like
limb.

# CHAPTER X.

THE ORDER URTICACEÆ : ILLUSTRATED BY THE COMMON NETTLE ;
THE HOP ; THE HEMP ; THE PELLITORY OF THE WALL ; THE
BREAD-FRUIT TREE ; THE JACK-TREE ; THE COW-TREE OR
PALO DE VACCA ; THE UPAS OR POISON-TREE OF JAVA ; THE
MULBERRY ; THE PAPER MULBERRY ; THE OSAGE ORANGE, OR
MACLURA ; THE COMMON FIG ; FICUS SYCAMORUS ; THE BAN-
YAN TREE ; THE INDIAN-RUBBER TREE ; AND FICUS RELIGIOSA.

THIS very large order is divided into two distinct tribes, which many botanists make separate
orders ; the one embracing the herbaceous
species with watery juice, and the other the
ligneous species, all of which have their juice
milky. The botanical construction of the
flowers is, however, strikingly alike in all, from
the nettle and the humble pellitory of the wall,
to the fig and bread-fruit tree. In all the
genera, the male and female flowers are distinct,
that is to say, some of the flowers have only
stamens, and the others only a pistil; the latter,
of course, being the only ones that produce
seed. None of the flowers have any corolla ;
and in all the male flowers, the stamens, which
are erect at first, spring back with elasticity to
discharge their pollen, and afterwards remain
extended. The seeds of all are enclosed in nuts:
though the eatable part varies, being in some

the dilated receptacle, as in the Bread-fruit and the Fig, and in others the metamorphosed calyx, as in the Mulberry. Many of the genera have one or two species which produce eatable fruit, though the fruit of the other species of the same genus is unwholesome; an anomaly rarely to be met with in any other order except Solanaceæ; and though the milky juice of most of the plants is poisonous, it affords in one species, the Cow-tree, wholesome food.

---

### TRIBE I.—URTICACEÆ.

ALL the plants contained in this tribe agree with the common Nettle in yielding a watery juice when broken; in their flowers having no corolla; in the male and female flowers being distinct; in the stamens being first erect, but springing back when they discharge their pollen, and remaining extended; and in their fruit being a nut. Most of them also agree in having rough leaves and angular stalks, the fibres of which are so tenacious as to be capable of being spun.

The common Nettle (*Urtica dioica*) is the type of this division; and we are so accustomed to consider it a noxious weed, that few persons are aware of the elegance of its flowers, which are disposed in drooping panicles. The male

flowers have their calyx divided into four sepals ;
and they have four stamens, the anthers of
which open with elasticity, and when they
spring back, the pollen, which is very abundant,
is discharged with such force that it may be
seen on a fine day in summer rising like a mist
or light cloud over the plants.  The stamens,
after they have discharged their pollen, lie extended and curved back over the segments of
the calyx, as shown at *a* in *fig.* 72.  The female

flowers have only two segments
to the calyx.  They have no
style, and the stigma, when
highly magnified, will be found
divided into numerous segments,
as shown at *b ;* the seed-vessel
is a nut, which has a shell and
kernel, the latter being the
seed.  The leaves are simple,

FIG. 72.—NETTLE
(*Urtica dioica*).

cordate, opposite to each other, and furnished
with stipules.  They are rough on the surface,
and covered with glandular hairs or stings.
These hairs are hollow, with a cell at the base
filled with a peculiarly acrid liquid, and tapering upwards so as to form a narrow tube, ending
in a sharp point.  When the point of the sting
enters the skin, the pressure compresses the cell
at its base, and the liquid it contains is forced
up the tube and injected into the wound.  The

stem is quadrangular, and its fibres are so
tough, that when separated from the pulp by
maceration, they may be spun into yarn.   The
young shoots when boiled are very good to eat.
The Roman Nettle (*U. pilulifera*) differs from
the common kind in having the male flowers in
loose panicles, and the female ones in compact
pill-like heads, whence the specific name.   The
sting of this nettle is worse than that of the
common kind.

The Hop (*Humulus Lupulus*) is a very interesting plant to a botanist, from the peculiarity
of its flowers.   The male and female ones are
distinct, and generally on different plants.   The

FIG. 73.—HOP (*Humulus Lupulus*).

male flowers are produced in loose panicles;
the calyx (*fig.* 73, *a*) consists of five sepals, in

the centre of which are five stamens, standing
at first erect, but springing back with elasticity,
when they discharge their pollen, and remaining
extended as shown at *b*.  The anthers open by
pores at the extremity of the cells, as in Ericaceæ.  The female flowers are produced in
close heads (*c*).  They have neither calyx nor
corolla, but the ovary of each is protected by a
membranous scale.  Each ovary has two styles,
though it produces only a single seed.  As the
fruit ripens the styles disappear, and the scales
enlarge, so as to give the head of female
flowers the form of a strobile or cone (*d*), the
ripe fruit or nut being placed at the base of
each scale, as shown at *e*.  The surface of the
scales is studded over with roundish glands,
which are filled with a substance resembling
pollen, called lupuline, which they give out on
pressure, as shown at *f;* and this substance
consists of a number of cells filled with volatile
oil, which occasion the fragrance of the hop, and
contain the bitter and astringent principles
which make the hop so useful in compounding
malt liquor.  The lupuline is also somewhat
narcotic; but though the fragrance of hops is
said to produce sleep when inhaled in small
quantities, an excess of it produces headache
and vertigo, especially in nervous persons.  The
leaves are opposite, and three or five lobed;

M

they are serrated on the edges, and rough on
the surface. The stems are angular, covered
with small prickles, and twining from left to
right. The fibres of the stem when separated
by soaking in water, are found to possess the
same kind of tenacity as those of the Nettle and
the Hemp, and may be made into cloth. The
young shoots when boiled, are very good to eat
as a substitute for asparagus. The leaves are
furnished with stipules, and the flowers spring
from the axils of the leaves.

The Hemp (*Cannabis sativa*), is an annual.
The male and female flowers are on different
plants as in the Hop and the Nettle. The male
flowers are produced in panicles, and the female
ones in heads separated by bracts, as shown in
a magnified female flower at *a* in *fig.* 74. The

FIG. 74.—HEMP.
(*Cannabis sativa.*)

ripe fruit or nut is enveloped in
a scale as shown at *b ;* and *c* is
a highly magnified section of
the nut. The male flower has
five stamens, and a calyx of
five sepals. The leaves are
opposite or alternate, and digi-
tate, that is cut into five long
segments like fingers, though
the upper leaves have only three
segments. They are serrated on the margin,
and rough on the surface. The fibres of the

stem, when separated from the pulpy part by
maceration, are manufactured into cordage;
and the seeds are mucilaginous, and are used
for feeding birds. The smell of hemp when
growing, produces the same effects as that of
hops in excess; and in hot countries it is followed by a kind of stupor, like that which is
the effect of opium.

The Pellitory of the wall (*Parietaria officinalis*), has the male and female flowers on the
same plant. The male flowers have four stamens, which spring back in the same manner as
those of the nettle; and the female flowers
have the same kind of stigma.

---

### TRIBE II.—ARTOCARPÆ.

THE plants included in this division differ so
widely in their general appearance from those
of the former tribe, that it is necessary to be a
botanist to perceive the resemblance between
them. When, however, they are botanically
examined, they will be found to agree in almost
every respect, except in their juice being milky
and glutinous intead of watery. The tribe
takes its name from the Bread-fruit tree (*Artocarpus incisa*). In this plant, the male flowers
are densely crowded round a spongy receptacle,
so as to form a long, somewhat club-shaped cat

kin.   Taken singly, each male flower consists of
a calyx divided into two sepals, and containing
a single stamen, with a two-celled anther, and
a very broad filament.   The female flowers are
placed round a globular receptacle, also of a
spongy consistency ;  and each consists of an
undivided calyx, hollow at the base to contain
the seed, and terminating in two styles.   The
styles wither as the seeds gradually ripen, but
the peaks of the female flowers remain, and
render the surface of the fruit rough.   The
fruit itself is the spongy receptacle, which gradually dilates and becomes more pulpy, till it
attains a very large size.   The greater part of
the ovules prove abortive, but those that ripen
retain their calyx, though they remain embedded
in the pulp.   The proportion of ripe seeds is
very small compared to the size of the eatable
part of the bread-fruit ;  frequently only four or
six seeds are found in a globe eight inches in
diameter ;  and many fruits produce no seeds at
all.   One variety, in particular, is always without seeds.   The fruit, when used, is generally
put into an oven or before a fire, and when the
rind turns black, it is scraped off, and the pulp
is found to resemble the crumb of new bread.
The seedless fruits are considered the best to
eat, and they are known by the smoothness of
their outer surface.   It adds to the interest ex-

cited by this singular tree, to recollect that the
Bounty, rendered so celebrated by the mutiny
of Christian, was sent out, under Captain Bligh,
to convey a number of plants of this tree from
Otaheite to the British settlements in the West
Indies; and that there actually were seven
hundred and seventy-four plants on board, at the
very time the mutiny broke out. The leaves of
the Bread-fruit tree are very large, being sometimes two or even three feet long, and a foot
and a half broad; they are leathery, and are
cut into from three to nine deep lobes. Their
colour is a deep green, with yellowish veins.
The petioles are short and thick, and there are
large stipules which wither and fall off before
the leaves. The whole plant abounds in milky
juice, which flows abundantly when the leaves
or branches are wounded or broken.

The Jack tree (*Artocarpus integrifolia*), bears
fruit of an oblong form often seventy or eighty
pounds in weight, the pulp of which is seldom
eaten; but the seeds, which are abundant, are
considered very good, and are said when roasted
to have the flavour of sweet chesnuts. The
leaves are very thick and leathery, and much
smaller than those of the Bread-fruit, being
seldom more than six or eight inches long.
They are also generally entire, but this is by no
means a constant character, notwithstanding

the specific name, as those near the root are
sometimes found nearly as deeply lobed as those
of *A. incisa*. The Jack tree is a native of the
East Indies, particularly of the Molucca Isles,
Amboyna, and Ceylon, and it also seems naturalised in the West Indies, particularly in the
Island of St. Vincent. The wood resembles
that of mahogany.

The Cow tree, or Palo de Vacca (*Galacto-
dendron utile*), appears nearly allied to the
Bread-fruit tree, though its flowers are un-
known. The nut, however, which is covered
with a husk apparently composed of the hardened calyx, resembles those of the other plants
belonging to the Urticaceæ, and the bark when
wounded gives out abundance of milk, which is
good to drink. Humboldt in his *Relation His-
torique*, describes this tree as "growing on the
sides of the rocks, its thick roots scarcely penetrating the stony soil, and unmoistened during
many months of the year by a drop of rain or
dew. But dry and dead as the branches appear," Humboldt continues, "if you pierce the
trunk, a sweet and nutritive milk flows forth,
which is in the greatest profusion at day-break.
At this time the blacks, and other natives of the
neighbourhood, hasten from all quarters, furnished with large jugs to catch the milk, which
thickens and turns yellow on the surface. Some

drink it on the spot, others carry it home to
their children ; and you might fancy you saw
the family of a cowherd gathering around him,
and receiving from him the produce of his kine."
(*Humboldt, as quoted in the Botanical Magazine,*
vol. 66, t. 3724.)

The Upas, or Poison tree of Java (*Antiaris
toxicaria*), about which so many fabulous stories
have been told, belongs to this tribe. The male
flowers are gathered together in small heads on
a fleshy receptacle, (see *fig.* 75 *a;*) and each
consists of a calyx of four
sepals (*b*), bending over four
stamens, with long anthers
and very short filaments.
The female flowers have an
undivided fleshy calyx with
two styles, and this fleshy
covering forms the pericardium of the fruit, which is a
drupe. When ripe, the fruit
represents a moderately sized
plum, inclosing the nut, or

FIG. 75.—UPAS TREE.

stone, which contains the kernel or seed. The
poison lies in the milky sap.

The common black Mulberry (*Morus nigra*)
has the general features of the order. The
male flowers grow together in a dense spike, as
shown in *fig.* 76 at *a*, and each flower consists

of a calyx of four sepals, and four stamens, which spring back and remain extended after

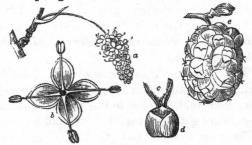

FIG. 76.—MULBERRY.

they have discharged their pollen (*b*). The female flowers also grow closely together, in dense spikes, round a slender receptacle; each having two elongated fringed stigmas (*c*), and a calyx of four sepals, and being inclosed in an involucre, as shown at *d*. As the seeds ripen, each female flower becomes a drupe, consisting of a fleshy and juicy pericardium formed from the calyx, and the nut; and these drupes being pressed closely together by the position of the female flowers, the whole adhere together and form the fruit we call the mulberry. The involucre withers when the calyx becomes juicy; but the remains of it and of the style are often seen on the ripe fruit, as shown at *e*. The receptacle also remains as a sort of core, which is thrown away when the fruit is eaten, though

it does not part from it so freely as in the raspberry; and the little nuts, or seeds as they are
called, are found in the centre of each juicy
globule.    The leaves are simple, entire, and
rough on the surface.

The white Mulberry (*Morus alba*) differs from
the common kind in the fruit not being eatable;
as the calyxes of the female flowers never become juicy.    The leaves are, however, much
smoother and of finer texture than those of the
black mulberry, and they are principally used
for feeding silkworms, for which those of the
black mulberry are not so good.

The red Mulberry (*M. rubra*) is an American
species, with leaves too rough to be good for
silkworms, and very indifferent fruit.    The Constantinople and Tartarian Mulberries are supposed to be only varieties of *M. alba*, though
their fruit is good to eat, and the latter has
lobed leaves.

The Paper Mulberry (*Broussonetia papyri-
fera*) has the male and female flowers on different plants.    The male flowers are produced in
pendulous catkins, and the calyx has a short
tube before it divides into four segments; each
flower is also furnished with a bract, but in
other respects their construction is the same as
that of the other flowers of the order.    The
female flowers have also a tubular calyx, and

they are disposed in globular heads on rather
long peduncles; but they differ from those of
the other genera in having only a single stigma,
and in the ovary being inclosed in an integument
within the calyx, which becomes juicy as the
seeds ripen, and not the calyx itself.   The
leaves are very irregularly lobed, and hairy;
and the liber or inner bark is used for making
what is called Indian paper.

   The Osage Orange (*Maclura aurantiaca*), has
the male and female flowers on different plants,
the male being borne in short close panicles of
ten or twelve flowers each, and not differing in
construction from those of the other genera.
The female flowers are borne on a large globular
receptacle, like that of the bread-fruit; and
they resemble those of that plant in construction, except that they are pitcher-shaped instead
of being angular, and that they have only one
stigma instead of two.   The receptacle also
never becomes soft and pulpy like that of the
bread-fruit, but remains hard and stringy and
unfit to eat.   The leaves are smooth and of
delicate texture, and as they abound in glutinous milk, they have been found very suitable
for silkworms.   The wood is of a beautiful
glossy texture, and very fine and close-grained.
The tree is found wild in the country of the
Osage Indians, near the Mississippi, and from

the rough surface of its fruit, and its golden-
yellow colour, it has received the name of the
Osage Orange.

The common Fig (*Ficus Carica*) has its male
and female flowers on the same plant, and often
within the same receptacle. The receptacle in
this plant instead of being surrounded by the
flowers, incloses them, and is, in fact, the fruit
we call a fig. This receptable is sometimes
roundish, but more generally pear-shaped; and
it is not quite closed, but has a little opening or
eye at the upper end, which is fitted in with
several very small scales. The stalk of the fig
is articulated on the branch. The male flowers
are generally in the upper part of the fig, and
they consist of a half tubular calyx, with a limb
divided into three segments, and three stamens.
The female flowers have each a calyx of five sepals,
and a single style with two stigmas; and they
are succeeded by the seeds, or nuts as they are
called, as each contains a kernel which is the true
seed. The leaves are very small when they first
expand, but they gradually increase in size, till
they become very large. They are generally lobed,
and their petioles are articulated. The figs are
produced in the axils of the leaves. It may be
observed here, that Du Hamel mentions that the
receptacle is not closed in all the varieties of
the fig, but that in some it opens naturally, when

the seeds are ripe, dividing at the orifice into four equal parts, like the valves of a capsule; and even when this is not the case, the figs, when the receptacle becomes pulpy and soft from ripeness, crack and burst at the sides, so as to allow of the escape of the seeds.

As the fig is not fit to eat till the seeds are ripe, various expedients have been devised to transmit the pollen from the male flowers which lie near the opening or eye, to the female flowers which lie nearer the stalk. In Italy this is called caprification, and is done by insects; but in the neighbourhood of Paris, a very small quantity of oil is dropped on the eye of the fruit as soon as it has nearly attained its full size.

There are several species of Ficus, though none of them will bear the open air in England except the common kind; and only two produce eatable fruit; viz., *F. Carica*, and *F. Sycamorus*, —the Sycamore tree of Holy Writ, which produces its small roundish fruit in clusters on the trunk and old branches, and not on the young wood, as is always the case with the common fig.

The other most remarkable species are the Banyan tree (*F. indica*), the figs of which grow in pairs, and are about the size and colour of a cherry; and the branches of which send down roots, which soon become equal in size to the

parent trunk, so that one tree soon becomes like
a small forest; the Indian-rubber tree (*F.
elastica*), the milky juice of which hardens into
Caoutchouc, though this substance is also produced by other trees, particularly by the Brazilian tree *Siphonia elastica;* and the Pippul
tree (*F. religiosa*).  The leaves of this last tree
are used in India for feeding silkworms, and it
is said that this is one cause of the strong and
wiry nature of the Indian silk ; and the insect
(*Coccus ficus*) feeds upon it and *F. elastica*,
which produces the substance called lac, of
which sealing-wax is made.  This species takes
its specific name of *religiosa*, from the legend
that the Hindoo god Vishnoo was born under
its branches.

# CHAPTER XI.

THE CATKIN-BEARING TREES : ILLUSTRATED BY THE WALNUT,
THE HICKORY, THE WILLOW, THE POPLAR, THE ALDER, THE
BIRCH, THE OAK, THE BEECH, THE SWEET CHESTNUT, THE
HAZEL, THE HORNBEAM, THE HOP HORNBEAM, THE PLANE
TREES, THE LIQUIDAMBAR, MYRICA, COMPTONIA, CASUARINA,
AND *Garrya elliptica.*

THE plants contained in this chapter are
placed by modern botanists in six or seven
different orders ; but I have been induced to
group them together, both because they follow
each other in regular succession, and because
there is a certain degree of general resemblance
which connects them together, and renders it
easier to retain their names when linked together
by the association of ideas, than it would have
been if they had been each described separately.

The first order of catkin-bearing trees that I
shall describe is called Juglandaceæ, and it contains three genera, only two of which, the
Walnuts and the Hickories, are common in
British gardens. The second order, Salicaceæ,
contains also two genera, the Willows and the
Poplars ; the third, Betulaceæ, contains both
the Alders and the Birch trees ; the fourth,

Corylaceæ or Cupuliferæ, contains the Oak, the
Beech, the sweet Chestnut, the Hazel, and the
Hornbeam; the fifth, Platanaceæ, is generally
considered to include two genera; viz., Platanus
and Liquidambar, though this last is, by some
botanists, placed in a separate order called
Balsamaceæ; and the sixth, Myricaceæ, or the
sweet Gale family. All the genera included in
these orders, with the exception of those belonging to Juglandaceæ, were formerly comprised
in one order, which was called Amentaceæ;
from the word Amentum, which signifies a catkin.
The seventh and last order I have mentioned
in this chapter is called Garryaceæ, and consists
of one single genus, Garrya, only lately known
in Europe. Of all these orders the largest and
most important is Cupuliferæ, as it includes,
among other valuable trees, the Oak and the
Beech. All the plants mentioned in this chapter
have their male and female flowers distinct,
many of the genera having them on different
plants; and the male flowers are always in catkins, generally long and cylindrical, but sometimes round and ball-shaped. The female
flowers are sometimes in catkins also, but sometimes they are produced singly or in pairs.
The flowers of both kinds are without petals, or
with such as are inconspicuous; and sometimes
without even a calyx, but they are always fur-

nished with bracts, which grow so closely to
the flower as almost to seem a part of it.   The
ovaries are generally two-celled, but they rarely
remain so, as they become one-celled before the
seed is ripe.   The style is, in most cases, very
short, and the stigma generally two-lobed.   The
leaves are always alternate, and generally simple,
except in the case of the Juglandaceæ.   They
are all hardy trees and shrubs.

JUGLANDACEÆ.—THE WALNUT TRIBE.

THE genera belonging to this order have compound leaves, and the male flowers in long
cylindrical catkins ;  the male and female flowers
being on the same plant.

### THE GENUS JUGLANS.

This genus consists of only three species :  the
common Walnut (*J. regia*) ;  the black Walnut
(*J. nigra*) ;  and the Butternut (*J. cinerea* or
*cathartica*).   The male and female flowers are
distinct, but on the same plant :  the male flowers
being produced in long, solitary, cylindrical catkins, and the female ones in pairs, or in shorter
catkins.   The leaves are pinnate, with the leaflets
not always opposite, which is very rarely the
case in other plants.   In *Juglans regia* (the
common walnut), the male flowers are produced
in a very thick catkin, each flower consisting of

a calyx divided into five or six scale-like lobes,
and generally from twelve to twenty stamens,
with very long anthers and very short filaments;
there is also a very curious bract to each, as shown
in the magnified
flower at *a* in *fig.*
77; in which the
anthers are seen
at *b*. The female
flowers are in
pairs, as shown at
*c*; and they con-

FIG. 77.—WALNUT (*Juglans regia*).

sist of a calyx, *d*, enclosing the ovary, and toothed
in the upper part, and four small petals encircling
two large thick leafy-looking stigmas, *e*.

The fruit is a fleshy husk in one piece, formed
of the dilated calyx; it generally retains the
stigmas till it has nearly attained its full size,
and when it becomes ripe it does not separate
into valves, but bursts irregularly. The nut,
on the contrary, is in two distinct valves, which
may be easily separated from each other; and
it is imperfectly divided into cells by four half
dissepiments. The germ of the future plant is
what children call the heart, and it is in the
upper part of the kernel, with the root end
uppermost, so that when a walnut is sown the
sharp end should be placed downwards. The
kernel is four-lobed, and deeply wrinkled; and

N

when the young plant begins to grow, it divides into two cotyledons or seed-leaves, which drop off when the true leaves are fully developed. The kernel is covered with a thick skin, which is very astringent; and the nut is covered with a membranaceous network of strong veins, which are generally found in a withered state on opening the ripe husk, having left their impression deeply imprinted on the outside of the shell of the walnut. The leaves are impari-pinnate, consisting of four pairs of leaflets and a terminal single one; the lower pair of leaflets is much the smallest, and the other leaflets are frequently not opposite; and they are sometimes unequal at the base. The main petiole is dilated at the point where it joins the stem; and the leaves are placed alternately. The tree is large and widely spreading; and the timber is of a close grain, and takes a fine polish.

The Black Walnut (*J. nigra*) differs from the

FIG. 78.—LEAF AND FRUIT OF THE BLACK WALNUT (*Juglans nigra*).

common kind, in the male flowers being on a smaller and more slender catkin, and furnished with a brown roundish bract at the back of the calyx. The female flowers are also in a sort of catkin, and four or five together. The fruit (see *a* in *fig.* 78) is round, and the husk very thick at first, but it gradually wastes away, when the seed is ripe, instead of opening. The leaves have seven or more pairs of leaflets, which are generally nearly opposite, and sometimes they are without the terminal single leaflet, as shown at *b*. The shell of the nut is very hard, and the dissepiments, which are also very hard, are generally perfect, and divide the kernel into four parts. The nuts should be sown as soon as possible after they are ripe, as they will not keep good above six months. The tree grows above seventy feet high, and the wood is very hard and black.

The Butter-nut (*J. cathartica*) resembles the Common Walnut in its male catkins, except that they are produced upon the old wood instead of being on the wood of the present year. The female flowers grow four or five together in a short catkin, and they are distinguished by their stigmas, which are rose-coloured. The fruit (*a* in *fig.* 79) is pear-shaped, ending in a rather long point; and the kernel of the nut (*b*) resembles that of the Common Walnut, except in being more oily. The

FIG. 79.—BUTTER-NUT (*Juglans cathartica*).

leaves (*c*) are like those of the Black Walnut, except that the leaflets are rather downy, and

N 2

that there is a terminal one.   The calyx of the
female flower is also covered with a viscid down,
which remains on the husk of the ripe fruit; and
the shell of the nut is very hard and very much
furrowed.   The tree is of much smaller size
than that of the Black Walnut, and it may
be easily distinguished by the greyness of the
bark of its young shoots; it also comes into
leaf earlier, and the nuts are ripe about a fortnight sooner than the others. The wood is light,
of a reddish colour, and rather a coarse grain.

### THE GENUS CARYA.

THE genus Carya (the Hickory) consists of
ten or twelve species, which greatly resemble

the Walnuts in their
general appearance, but
are distinguished by the
male catkins, instead of
being solitary, being produced in tufts or bunches,
three or more on each
peduncle.   The stigma
is also frequently four-
lobed, and the husk,
when ripe, divides into

FIG. 80.—THICK-SHELL BARK
HICKORY (*Carya lacinosa*).

four equal valves, which in some of the species are
very thick, as in the Thick-shell bark Hickory
(see *a* in *fig.* 80). The nut (*b*) is not valved, and it

is either not furrowed, or very slightly so; but it
has four angles which are more or less distinct
in the different species: the shell and the dissepiments are both very hard, and the latter, as in
the Mocker nut, are sometimes entire, so as to
render it very difficult to extract the kernel.
The leaves (c) resemble those of the walnut;
but they are generally of a thinner texture,
and somewhat downy, the down being disposed
in little tufts, as may be seen by a microscope.
The trees vary much in size, but all of them
have a reticulated bark. The wood is of a coarse
grain, and will not polish ; but it is very strong,
and so remarkably tough that it is hardly possible
to break it.

There is only one other genus in the order
Juglandaceæ, and that consists of only a single
species, *Pterocarya caucasica*. It has pinnate
leaves of nineteen leaflets each, placed as closely
as possible together; and the fruit, (that is, the
husk,) is spread out on each side into a thin
membrane or wing. This plant is sometimes
called *Juglans fraxinifolia*.

---

## SALICACEÆ.—THE WILLOW TRIBE.

THE plants contained in this order have
simple leaves, and the male and female flowers

on different plants, both in upright cylindrical
catkins.

### THE GENUS SALIX.

THE genus Salix (the Willow) contains perhaps more species than any other, above two
hundred and fifty having been named and described, besides innumerable varieties. The plants
included in the genus may, however, be all
divided into three kinds—viz. the true Willows,
which have thin green leaves, and which include
all the tree species, most of which have brittle
branches ; the Osiers, the leaves of which resemble those of the Willows, but which are low
shrubs with very tough branches ; and the
Sallows, the leaves of which are thick and woolly
or shaggy. The Osiers and the true Willows
are often confounded together; particularly when
the former take, as they sometimes do, a tree-
like character ; but the Sallows are always perfectly distinct. The rods of the Osiers are used
in basket-making.

All the species of the genus have their male
and female flowers on different plants, both
kinds of flowers being placed on short catkins
which are either erect or spreading sideways.
The male flowers have each from one to five or
more stamens, with no petals or calyx, but as a
substitute a bract or scale, which is entire and
hairy, and which has one or more glands at its

base.   The female flower has a similar bract or
scale and it is also without either petals or calyx;
there are two stigmas, each of which is sometimes two-lobed.   The capsule has only one
cell, but many seeds which are covered with
down or longish hairs, and which are very
conspicuous from the capsule opening naturally
into two valves when ripe.   The leaves of the
Osiers and Willows are generally lanceolate,
and serrated at the margin, and they are
always furnished with stipules ; but the leaves
of the Sallows are generally much broader, and
sometimes roundish ; and they are always of a
thick velvety texture.   Though the number of
the stamens varies in the different species, two
are by far the most common.

   *Fig.* 81 shows the female flower of
*Salix fragilis* at *a*, *b* is the honey gland,
*c* the stigma, which is divided into four
equal parts, and *d* the bract or scale
with its hairy fringe ; *e* is the male
flower with its two stamens, two glands,
and hairy scale. This species is a tall,
bushy-headed tree, with the branches
crossing each other frequently, being
set on obliquely ; and it is called
the Crack Willow, from the young
branches separating from the trunk in
spring with the slightest blow or jerk, their

FIG. 81.—THE
WILLOW
(*Salix*).

bases being as brittle as glass.  The leaves are
of a deep green.  The White Willow (*Salix
alba*) differs from the preceding species in the
branches being widely spreading and somewhat
drooping, the old bark cracked into deep
fissures, and the foliage of a silvery grey, owing
to the silky hairs with which the leaves are more
or less covered.  The wood of the Tree Willows
is soft and white, and very elastic; it is therefore used for cricket-bats, mallets, and other
purposes where wood is wanted to resist a hard
blow.  *S. vitellina*, the Golden Osier, is so called
from its golden-coloured bark ; and *S. purpurea*,
the Purple Willow, is so called from the colour
of its branches.  This last species has only one
stamen; but as the anther is four-celled, it is probably two stamens grown together.  All the
species that have only one stamen have a four-
celled anther, as for example the Rose Willow
(*S. Helix*), which has the female catkins red.
*Salix caprea*, the great round-leaved Sallow or
Palm Willow, is perhaps the handsomest species,
from the great abundance and golden hue of its
flowers.

### THE GENUS POPULUS.

THE genus Populus (the Poplar) is distinguished from Salix by the bracts of the flowers
being deeply cut instead of being entire ; by
both the male and female flowers having a calyx;

and by the male flowers never having less than
eight stamens.  The leaf-buds are also covered
with numerous scales.   *Fig.* 82, *a,* shows the

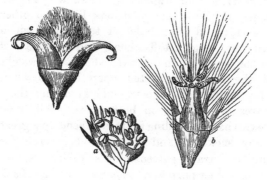

FIG. 82.—TREMBLING POPLAR OR ASPEN (*Populus tremula*).

stamens of the Trembling Poplar or Aspen
(*Populus tremula*) shrouded in their cup-like
calyx, and with their laciniated bract ; *b* shows
the female flower with its four stigmas and
deeply-cut bract ; and *c,* the pod with its valves
curling back, so as to show the downy covering
of the seeds.   All these parts are magnified to
show them distinctly, as they are nearly the
same in all the species.   The following are the
distinctions between the principal species.   In
the White Poplar, or Abele-tree (*P. alba*), the
leaves are lobed, and covered with a white down
on the under side.   In *P. canescens,* the Grey

Poplar, the leaves are also downy beneath; but
they are roundish, and the female flower has
eight stigmas instead of four. The Aspen (*P.
tremula*) has four stigmas, with two leafy appendages at the base, which look like two other
stigmas; and the petioles of the leaves, which
are very long, are flattened, and so attached
to the stem as to be twisted by the weight
of the leaf when acted upon by the wind,
which gives them their tremulous motion : these
leaves are smooth on both sides. All these
species have spreading roots, and send up a great
many suckers; and their wood is used for
butchers' trays, pattens, bowls, milk-pails, and
various other purposes. *Populus nigra*, the Black
English Poplar, on the contrary, does not send
up suckers, and its wood is of very little use;
it is, however, very ornamental from the large
size and great number of its male catkins, and
the bracts of the flowers being of a brownish
red, which gives them, when fallen, the appearance of the large brownish-red caterpillars of
the Goat-moth. The Black Italian Poplar (*P.
monilifera*) is remarkable for the quickness of
its growth. The capsules of the female trees
contain such a quantity of down attached to the
seeds, as to render it quite unpleasant to walk
under them when they are ripe. The Lombardy
Poplar (*P. fastigiata* or *dilatata*) is remarkable

for its upright and close habit of growth ; its leaves also are very peculiar in their shape, being broad at the base and then tapering suddenly to a point. The seeds resemble those of the Black Italian Poplar in the quantity of wool which they produce, but luckily the female plants are extremely rare. There are many other species, the most remarkable of which are the Carolina Poplar (*P. angulata*), known by its square stem and very large leaves ; the Balsam Poplar, or Tacamahac tree (*P. balsamifera*), the buds of which are covered with a resinous fragrant substance, and the leaves are of a pale yellowish green, appearing very early in spring ; and the Ontario Poplar (*P. candicans*), which resembles the balsam Poplar, except in its leaves, which are very large and whitish on the under surface, and in the great rapidity of its growth, while that of the Balsam Poplar is rather slow.

BETULACEÆ.—THE BIRCH TRIBE.

The plants included in this tribe have single leaves, which are generally what is called feather-nerved ; that is, the veins are marked strongly and deeply from the mid rib to the margin. The flowers are in cylindrical catkins, the male and female flowers being on the same plant.

## THE GENUS BETULA.

THE common Birch (*Betula alba*) is an exceedingly graceful tree. The male
catkins are produced singly, or two
or three together. They are long,
slender, loose, and gracefully drooping; (see *fig.* 83;) and each consists
of a great number of flowers, pressed
close together, and growing round a
rachis or stem, as shown in the

FIG. 83.—CAT-
KINS OF THE
BIRCH.

catkin *a* in *fig.* 84, from which some of the
flowers have been removed. The male flowers

have each ten or
twelve stamens
enclosed in three
or more scales or
bracts, as shown
in a reversed
flower at *b*. The
female flowers are
produced in dense
catkins, which
are much shorter
than the others,
and always soli-
tary; the flowers,
which are arrang-

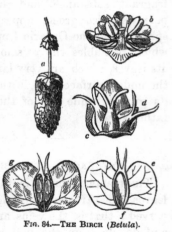

FIG. 84.—THE BIRCH (*Betula*).

ed round a very slender axis, are furnished with
lobed scales, and *c* is a scale with three female

flowers in its lobes, each having two long spreading stigmas (*d*). A ripe capsule is shown at *e*, with its membranaceous wings, and the cell *y* open to show the seed. The ovary when young has two cells and two ovules (as shown at *g*) ; but the division between the cells wastes away as the seeds ripen, and one of the ovules proves abortive. There are several species of Birch natives of America, some of which have upright oval female catkins like those of the Alder, but they are always distinguished by being solitary.

The bark of the Birch is remarkable for its tenacity, and for the great length of time that it will resist decay. In America they make canoes of the bark of *B. papyracea ;* and in Lapland huts are thatched with that of *B. alba.* The Birch is remarkably hardy ; and it grows nearer the limits of perpetual snow both on mountains and near the pole than any other tree.

### THE GENUS ALNUS.

THE Common Alder (*Alnus glutinosa*), though so nearly allied to the Birch botanically, differs widely in its habits ; as it always grows in low marshy situations, or near water, while the Birch prefers the summits of the loftiest hills. In the Alder, the male catkins are long and drooping, like those of the Birch ; but they are generally produced in clusters of three or more

together. The male flowers are furnished with
three lobed bracts or scales, each containing
three flowers, each flower having a calyx of
four scales united at the base, and bearing four
stamens. The female flowers are in close ovate
catkins, produced in clusters of four or five
together, instead of being cylindrical and solitary, as in the Birch; the scales of the catkins,
though three-lobed, are only two-flowered, and
the flowers have two long stigmas like those of
the Birch. The ovary has two cells and two
ovules, but it only produces one seed. The
ripe fruit is a nut without wings, attached at
the base to the scale of the cone-like catkin,
the scales of the catkin becoming rigid, and
opening, like those of the Scotch Pine, as the
seed ripens. There are several species of Alder,
some of which bear considerable resemblance
to the American species of Birch; but they are
easily distinguished by the female catkins of
the Birch being always solitary, while those of
the Alder are produced in clusters, and by the
capsules of the Alder being without wings.

---

### CUPULIFERÆ—THE CUP-BEARING TREES.

This order includes six genera of very important trees; all of which have their ripe fruit
shrouded in a cup-like involucre, which they

rctain till ripe.  The male and female flowers
are on the same plant.

### THE GENUS QUERCUS.

THE fruit of all the species of Oak is an
acorn, which is only partly covered by a scaly
involucre called the cup.  The shape of the
acorn, and the height to which it is covered by
the cup, differ in the different species ; but the
general character of both is always the same.

The male catkins of the common British
Oak (*Quercus Robur pedunculata*) are long and
very few flowered; the flowers being small and
very far apart.  The flowers themselves have
six or eight stamens and as many feathery
bracts, which are united at the base.  The
female flowers (*a* in *fig.* 85)
are produced on a long
stalk at a distance from
each other, and each con-
sists of an ovary closely
covered with a toothed
calyx, as shown in the
highly-magnified flower at
*c*, and an involucre of several
bracts or scales, *d* ; the

FIG. 85.—THE OAK.

style is short and thick, and the stigma (*e*) is
three-lobed.  As the fruit ripens, the style and
stigma wither away, and the seed remains

covered by the adnate calyx (*b*), which has
become hard and shining.  There is a circular

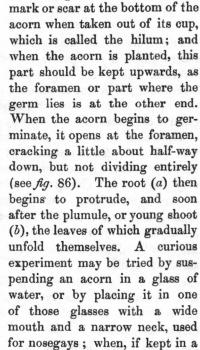

mark or scar at the bottom of the
acorn when taken out of its cup,
which is called the hilum; and
when the acorn is planted, this
part should be kept upwards, as
the foramen or part where the
germ lies is at the other end.
When the acorn begins to germinate, it opens at the foramen,
cracking a little about half-way
down, but not dividing entirely
(see *fig.* 86).  The root (*a*) then
begins to protrude, and soon
after the plumule, or young shoot
(*b*), the leaves of which gradually
unfold themselves.  A curious
experiment may be tried by sus-
pending an acorn in a glass of
water, or by placing it in one
of those glasses with a wide
mouth and a narrow neck, used
for nosegays ; when, if kept in a
sitting - room, the acorn will
gradually open, and the root and

Fig. 86.—Germina-
tion of the Acorn.

leaves develop themselves; and thus may be
watched the first beginning of the monarch of
the forest, the progress of which is so strikingly

depicted in the beautiful lines adapted by
Cowper to the hollow trunk of a gigantic oak in
Yardley Chase near Castle Ashby:—

Thou wert a bauble once, a cup and ball
Which babes might play with ; and the thievish jay,
Seeking her food, with ease might have purloin'd
The auburn nut that held thee, swallowing down
Thy yet close folded latitude of boughs,
And all thy embryo vastness, at a gulp.

&ast;        &ast;        &ast;        &ast;        &ast;        &ast;

Time made thee what thou wert—King of the Woods !
And time hath made thee what thou art—a cave
For owls to roost in !   Once thy spreading boughs
O'erhung the champaign, and the numerous flock
That grazed it, stood beneath that ample cope
Uncrowded, yet safe shelter'd from the storm.

&ast;        &ast;        &ast;        &ast;        &ast;        &ast;

Embowell'd now, and of thy ancient self
Possessing nought but the scoop'd rind, that seems
A huge throat calling to the clouds for drink,
Which it would give in rivulets to thy roots ;
Thou temptest none, but rather much forbidd'st
The feller's toil, which thou wouldst ill requite.
Yet is thy root sincere, sound as a rock :
A quarry of stout spurs and knotted fangs,
Which, crook'd into a thousand whimsies, clasp
The stubborn soil, and hold thee still erect.
Thine arms have left thee—winds have rent them off
Long since ; and rovers of the forest wild
With bow and shaft have burnt them.   Some have left
A splinter'd stump, bleach'd to a snowy white ;
And some, memorial none where once they grew.
Yet life still lingers in thee, and puts forth
Proof not contemptible of what she can,
Even when death predominates.   The spring

Finds thee not less alive to her sweet form,
Than yonder upstarts of the neighbouring wood,
So much thy juniors, who their birth received
Half a millennium since the date of thine.

The leaves of the common Oak are deeply
sinuated, and without footstalks, but those of
*Quercus Robur sessiliflora,* another British Oak,
are upon short footstalks, though the acorns
are sessile.  This last species predominated in
the oak forest which formerly surrounded London; and many examples are still to be found at
Lord Mansfield's beautiful seat at Hampstead,
the name of which, Ken wood, alludes to them,
Ken being Saxon for an acorn.  The wood of
this tree was also used for the roof of West-
minster Hall, and many other ancient buildings
which till lately were supposed to be of Chestnut.
Oak wood may always easily be tested by
wetting a knife and then cutting it, when the
astringent property in the Oak will turn the
knife black, a result that will not take place
with Chestnut.

There are nearly fifty species of Oaks which
may be obtained in the British nurseries; the
most remarkable of which are the Cork tree
(*Quercus Suber*), the cork being the bark; the
Evergreen Oak (*Q. Ilex*); the American Oaks,
particularly the scarlet Oaks (*Q. coccinea* and
*Q. rubra*), the Live Oak (*Q. virens*), and the
Willow Oak, with long narrow entire leaves

(*Q. Phellos*) ; and the Turkey, Fulham, and Lu-
combe Oaks (*Q. Cerris* and its varieties). All Oak
trees are very liable to be attacked by a species
of gnat, and which produces excrescences on the
branches.   The oak apples of the British Oak,
and the galls of *Quercus infectoria*, which are
used in making ink, are of this nature.   The
Kermes, an excrescence found on *Quercus cocci-
fera*, is the work of a kind of Coccus, similar
to that which produces the cochineal on the
Opuntia.

The timber of all the European Oaks is
remarkably durable ; but that of nearly all the
American Oaks, except *Quercus virens*, is coarse
grained, and so porous that it cannot be used
for wine casks.   The cork trees are generally
grown in Spain ; and as the cork when taken
off the tree, curves round, it is laid upon the
ground and kept flat with heavy stones ; while
a fire is made upon it with the branches, so as
to heat it through, after which it remains flat
when the stones are removed.

## THE GENUS FAGUS.

THE Beech (*Fagus sylvatica*) bears very little
resemblance to the Oak.   The male flowers are
in globular catkins (see *a* in *fig.* 87), each
flower consisting of a bell-shaped calyx (*b*), cleft
into five or six teeth, and containing eight or

o 2

ten stamens, which project beyond it. The
female flowers also grow in globular heads (c)

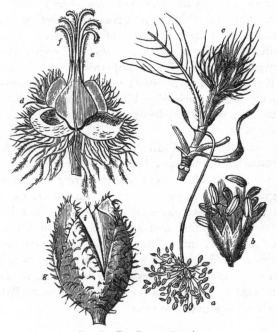

FIG. 87.—THE BEECH (*Fagus*).

two or three together, surrounded by a great
number of linear bracts, which gradually grow
together, and form a four-lobed involucre shown
open at *d*. In the centre of this involucre are
two or more female flowers, each surrounded by

a hairy calyx, cut into teeth at the tip (*e*). Each flower has three styles (*f*) ; and the ovary, which is sharply angular, has three cells, with two ovules in each. As the fruit swells, the linear bracts diminish, till at last they have only the appearance of small spines on the involucre (*g*), which opens when ripe into four valves (*h*), and contains two or three angular nuts (*i*), which are called the mast. The leaves of the Beech are of thin and delicate texture, and they are strongly feather-nerved. The tree is large and very handsome, and it is easily known, even in winter, by the smooth shining white bark of the main trunk. There are only two species of Fagus common in British gardens, and these are the common Beech (*Fagus sylvatica*) which has numerous varieties, including one with dark reddish purple leaves, generally called the Purple Beech ; and the American Beech (*F. ferruginea*), the leaves of which are copper-coloured.

There are, however, two species from Terra del Fuego, which have been introduced, but they are at present rare. One of these (*F. betuloides*) is called the Myrtle tree in Van Dieman's Land, where it is also found wild, and it is remarkable for producing a fungus on its trunk, which, when cut in slices and cooked, is said to be very good to eat.

## THE GENUS CASTANEA.

THIS is a very small genus, only containing two or three species, of which only one, the Sweet Chestnut (*Castanea vesca*) is common in England. This plant was included by Linnæus in the genus Fagus, but it appears very distinct. The male flowers are produced round a central axis, but so far apart as hardly to be like a catkin (see *a* in *fig.* 88). These flowers in the bud look like little knobs, but when they open the stamens burst out, as shown at *b*. Each flower has a large and a small bract, and from ten to fifteen stamens. The female flowers are disposed in a tuft as shown at *c*, surrounded by a number of bracts

FIG. 88.—CHESTNUT (*Castanea vesca*).

and scales, which afterwards grow together and form a spiny involucre (see *fig.* 89 *a*,) which forms the husk of the ripe nuts (*b*), and opens into four valves as shown at *c*. Each female flower has a closely-fitting calyx, toothed at the tip, which afterwards becomes the hard brown skin that envelops the kernel of the ripe nut ; and

each flower is furnished with six styles, having
as many cells with two ovules in each, though

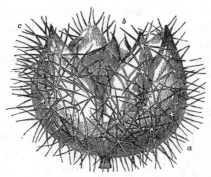

FIG. 89.—FRUIT OF THE SWEET CHESTNUT.

generally all the cells unite into one, and most
of the ovules wither before the fruit ripens.
There are three female flowers in each involucre,
which lie nestling together like birds in a
nest.   When ripe the involucre or husk opens
naturally into four valves (as shown in *fig.* 89),
and drops the one or two Chestnuts which it
contains.   Each nut, when ripe, is enveloped in
a brittle shining skin formed of the metamorphosed calyx, and consists of only one cell, in
which are one, two, or at most three kernels,
which are the seeds.

### THE GENUS CORYLUS.

THE Hazel Nut (*Corylus Avellana*) has the
male and female flowers on the same tree; the
male being in long catkins and the female ones
in little oval buds, something like those of the
Oak, (see *a* in *fig.* 90,) which are so small that
they would hardly be seen on the tree, if it
were not for their bright red stigmas. The

male flowers (*fig.*
91) have each three
bracts, one behind
the two others, to
the inner ones of
which are attached
eight stamens. As
the buds containing
the female flowers

FIG. 90.—THE HAZEL (*Corylus
Avellana*).

expand, two or three small leaves make their
appearance between the scales (*b* in *fig.* 90), so
that gradually the bud becomes a
little branch, bearing the female
flowers at its tip. Each flower has
two long stigmas, and the ovary is
enclosed in a closely-fitting calyx (*c*)
toothed at the upper part, the whole
being enveloped in a deeply cut in- FIG. 91.—MALE
volucre (*d*), which afterwards becomes FLOWER OF THE
                                          HAZEL.
the husk of the ripe nut. This in-

volucre is not closed, as in Fagus and Castanea,
but it is open at top ; the nut, as in all cupuliferæ, adhering to it, when young, by the hilum or
scar visible at its base.   There are about seven
species of Corylus, of which the most remarkable is the Constantinople Nut (*Corylus Colurna*).
The Filbert is only a variety of *Corylus Avellana*.

THE GENUS CARPINUS.

Some botanists include this genus and that
of Ostrya in the order Betulaceæ, instead of
placing them in Cupuliferæ, as the nut of the
Hornbeam is not surrounded by a cup or husk,
but by a leaf-like involucre as shown in *fig.* 92,
at *a, b* being the nut.
Both the male and the
female flowers are pro-
duced in long catkins,
which have an exceed-
ingly light and elegant
appearance on the tree.
The male flowers con-
sist each of one bract,

FIG. 92.—FRUIT OF THE HORN-
BEAM.

with twelve or more stamens attached to its
base ; and the female flowers have each two
very long stigmas, and a ribbed calyx, which
adheres to the ripe nut and assumes the appearance of a hard brown skin.   The leaves are
feather-nerved and persistent, like those of the

Beech, frequently remaining on the tree, though
in a withered state, till spring.

The nut appears ribbed when ripe, from the
remains of the metamorphosed calyx, and it
contains only one seed ; though, as in the other
allied genera, the ovary had two cells, with an
ovule in each.

### THE GENUS OSTRYA.

THE Hop Hornbeam (*Ostrya vulgaris*) was
included in the genus Carpinus by Linnæus ;
and indeed the general construction of the
flowers is the same. The male catkins are,
however, very much longer, and the female
catkins much shorter, and closely resembling
those of the Hop.

### THE ORDER PLATANACEÆ.

THIS order formerly included the Plane trees
and the Liquidambar ; but many botanists now
put the latter tree in a separate order, which
they call Balsamacea.

### THE GENUS PLATANUS.

IN the Oriental Plane (*Platanus orientalis*)
the male and female flowers are both in globular
catkins. The male flowers are composed of very
small, but rather fleshy bracts, which remain on
after the stamens fall; and the female flowers

are each furnished with bracts, and have two
long stigmas. Both kinds of flowers are so
small as not to be seen without a microscope.
The fruit is covered with fine hair. The
globular catkins retain the bracts, and these
remain on after the seed has fallen, giving the
tree a very singular appearance even in winter.
The Occidental Plane (*P. occidentalis*) differs
principally from the preceding species in the
leaves being more downy beneath; the buds
are also so downy that the tree in America is
called the Cotton-tree. Both kinds are remarkable for the manner in which the bark becomes
detached from the main trunk and peels off.

### THE GENUS LIQUIDAMBAR.

THE common Liquidambar (*L. styraciflua*) is
remarkable for the beautiful crimson which its
maple-like leaves take in autumn. The male
flowers are on an upright catkin, and the female
ones in a globular one, like the Planes. When
the fruit is ripe, the numerous capsules that
surround the globular catkin burst, and the
seeds, which are winged, are scattered by the
wind.

### THE ORDER MYRICACEÆ.

THE principal genera are Myrica, the Sweet
Gale ; Comptonia, a curious shrub with fern-

like leaves; and Casuarineæ, a New Holland
tree without leaves, but with jointed leaf-like
stems.

### THE GENUS MYRICA.

THE male flowers are produced in rather long
erect catkins, each having only one scale, and
four stamens. The female catkins are short,
and each flower has three scales or bracts; the
ovary has two long stigmas, and the fruit is a
drupe, the scales becoming fleshy when ripe.
The bracts and leaves are covered with glands
filled with aromatic oil; and in *M. cerifera*, the
fruit is covered with a waxy secretion, which is
used as wax.

---

### THE ORDER GARRYACEÆ.

THIS order consists of only one genus, Garrya.

### THE GENUS GARRYA.

*Garrya elliptica* is an evergreen shrub remarkable for its long and graceful male catkins,
the flowers of which consist of four stamens
within a four-cleft calyx, enclosed within bracts
united at the base. The female flowers are on
a different plant, and the fruit is a berry not
opening naturally.

# CHAPTER XII.

THE greater part of the trees included in this
chapter are comprised by Richard, De Candolle,
and other foreign botanists, in the order Coniferæ; which they have divided into three
sections: viz., the Abietineæ, or Pine and Fir
tribe; the Cupressineæ, or Cypress tribe; and
the Taxineæ, or Yew tribe. The last tribe,
however, Dr. Lindley has formed into a separate
order, which will probably be adopted. Most
of the genera have, what the Germans so graphi-
cally call needle leaves; that is, their leaves
are long and narrow, and terminate in a sharp
point. The flowers also are quite different from
what is generally understood by that name;
being in fact nothing but scales: those of the
male flowers containing the pollen in the body
of the scale, and those of the female producing
the ovules, or incipient seeds at the base. The
fruit of the Abietineæ is a cone, the scales of

which open when the seeds are ripe. That of the
Cupressineæ is also called a cone by botanists,
but it is rounder, and has not so many scales.
The fruit of the Taxineæ is an open succulent
cup, bearing the seed or nut in its centre.

Linnæus placed nearly all the hardy Abietineæ
in the genus Pinus, and since his time botanists
have disagreed exceedingly respecting the generic
names of the different plants; no less than
twelve different divisions of them having been
published, by as many eminent botanists, since
the commencement of the present century.
The best, however, appears to be that of M.
Richard, which was approved by De Candolle,
and which has been adopted with a slight altera-
tion in Mr. Loudon's *Arboretum Britannicum.*
According to this arrangement, the hardy
Abietineæ are divided into five genera; viz.,
Pinus, the Pine, including all the resinous trees
with long leaves, which grow two or more
together in a sheath; Abies, the Spruce Fir,
the leaves of which do not grow in a sheath,
but are scattered round the branches, the leaves
themselves being short, flat, and the same on
both sides; Picea, the Silver Fir, the leaves of
which resemble those of Abies, except that the
edges curl in, and the under surface is quite different from the upper one, being marked with
two white lines, one on each side the midrib; the

leaves are also placed nearly in two rows, one on each side the branch; Larix, the Larch, the leaves of which are very slender and produced in tufts, but which fall off every winter; and Cedrus, the Cedar, the leaves of which resemble those of the Larch, but which do not fall off every winter. The distinctions between these genera in the leaves only are very clear, and easily remembered; and their cones differ as decidedly: those of the Pines are hard and thick at the tips of the scales, which remain on after the seed drops; those of the Spruce Firs are thin at the tips of the scales, which also remain on the cones after they have lost their seeds, and the cones are drooping, and tapering at both ends; those of the Silver Firs are erect, cylindrical, and of nearly the same diameter throughout, and the scales fall with the seeds; those of the Larch are erect, but small and conical, and the scales remain on after the seeds have fallen; and those of the Cedar are erect, oval, and with deciduous scales. To the hardy genera may now be added Araucaria, as one species of this genus (*A. imbricata*) has been found quite hardy in Britain.

The Cupressineæ are divided into four or five genera; viz., *Thuja*, the Arbor Vitæ, some of the species of which have been formed into a new genus under the name of Callitris; *Cupressus*,

the Cypress; *Taxodium*, or *Schubertia*, the
deciduous Cypress; and *Juniperus*, the Juniper.
The only needle-leaved trees belonging to
Taxineæ belong to the genus Taxus, the Yew,
unless we separate from it the new genus
Torryea.

_____

§ 1. THE ABIETINEÆ.—THE PINE AND FIR TRIBE.

THE plants included in this section, with the
exception of the Larch, are evergreens.  They
are all lofty trees, with straight erect stems,
and their branches growing in whorls or tiers,
so as to produce a very peculiar and striking
effect.  The male and female catkins are on the
same plant; the female one containing two seeds
at the base of each scale.  The pollen of the
male flowers is so abundant that any one passing through a grove of these trees in May or
June, might fancy it was raining brimstone.
Most of the species are timber trees, producing
the wood called deal; that used for the flooring
and other parts of houses, being principally the
wood of the Scotch Pine, and the Norway
Spruce.  Most of the species produce turpentine, which is the thin part of the sap which
flows from the tree when a notch is cut in the
trunk; the thick part of the sap when purified
by boiling is the yellow resin.  Tar is produced
by cutting the roots and wood of pine and fir

trees into pieces, and putting them into a sort
of oven; when the tar runs from the charred
wood, and lamp-black is made from the soot
which collects on the roof of the oven. Pitch
is boiled tar. Pyroligneous acid is obtained by
burning the wood into charcoal in an iron
cylinder, and condensing the vapour that arises
from it.

### THE GENUS PINUS.

THIS genus, according to Linnæus, was made
to include all the Pines and Firs, the Cedar
and the Larch; and this arrangement has been
followed by the late A. B. Lambert, Esq., in his
magnificent work
on this tribe. In
its present re-
stricted form, it
contains only those
plants that have
long slender leaves,
which are pro-
duced in membra-
naceous sheaths,
(see *a* in *fig.* 93)
two, three, or five
together. The
male flowers are
produced in long

FIG. 93.—BRANCH OF THE SCOTCH PINE.

upright catkins, (*m*) growing two or three toge-

P

ther, and they consist each of one scale, which is
surmounted by a kind of crest, *b*. The pollen is
contained in two cells formed in the body of each
scale, which open lengthways, as shown in the
scale of the Scotch Pine (*Pinus sylvestris*), at *g* in
*fig*. 94. The female scales or carpels when ripe

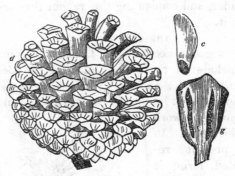

FIG. 94.—CONE OF THE SCOTCH PINE (*Pinus sylvestris*).

form a strobile or cone (*d*), and in the Scotch
Pine they are thickened at the tip (*e* in *fig*. 93);
but when young they appear as shown at *f*.
Each scale is furnished with a thin membrane-
like bract on the outside, which is conspicuous
when young, but which is hidden by the scales
in the ripe cone; and each has two seeds inside,
which are each furnished with a long thin transparent wing (*c*). When the seed is ripe, the
cone opens as shown at *d*, and the seeds falling
out are carried away by the wind. When the

seed is sown and begins to germinate, the young plant sends down a root, and pushes through the ground its upright shoot, which has six cotyledons, bearing the husk of the seed upon their tip. All the species of the genus Pinus agree with the Scotch Pine in the construction of their flowers, and they differ from each other principally in their cones, and in the number of leaves which they have in a sheath. By far the greater number have two leaves in a sheath, (see *a* in *fig.* 93,) and among these are the Scotch Pine (*P. sylvestris*), which has small straight cones without prickles ; *P. Banksiana,* which has crooked cones ; *P. pungens* and other American Pines, which have prickly cones, every scale being furnished with a sharp spine ; the Corsican Pine (*P. Laricio*), and several allied species, which have no spines on their cones, but every scale curving outwards ; the Pinaster (*P. Pinaster*) which has large cones, with very short broad spines, and the Stone Pine (*P. Pinea*), the cones of which are smooth and shining, and very large, and the seeds of which are eaten. The pines that have three leaves in a sheath, are chiefly natives of North America, and have prickly cones ; such as *Pinus Tæda* and its allies, *P. ponderosa,* remarkable for its heavy wood which sinks in water, and its large spreading branches ; and

*P. Sabni* and *P. macrocarpa*, which have long,
slender, drooping
leaves, and very
large hooked cones.
The pines which
have five leaves in
a sheath, include,
among others, the
Weymouth Pine
(*P. Strobus*), the
cones of which are
long, narrow, and
drooping (see *fig.*
95); *P. Lamber-*
*tiana*, which has

Fig. 95.—Weymouth Pine (*P. strobus*).

cones above a foot long; and *P. Cembra*, which
has an oval cone, the scales of which are concave, and the seeds without wings.

### THE GENUS ABIES.—THE SPRUCE FIR.

This genus includes all the Spruce Firs, and
they are readily distin-
guished from the pines by
their drooping cones (see
*a* in *fig.* 96), the scales of
which are not thickened at
the tip, but drawn out into
a thin brittle membrane;
and their leaves, which do
not grow erect in sheaths,

Fig. 96.—Spruce Fir (*Abies*
*excelsa*).

but in rows standing out from the branches (*b*), and which being the same on both sides, look as if two had grown together to make one. The difference between the Pines and the Firs will be seen clearly by comparing *fig.* 96, which represents a branch of the Spruce Fir, with *fig.* 97, which represents a branch of *Pinus pumilio*, a dwarf variety of the Scotch Pine.

FIG. 97.—*Pinus pumilio.*

The common Spruce Fir (*Abies excelsa*) is a tree of stately growth, with an erect pyramidal form, and numerous tiers of drooping branches.  It is the loftiest of European trees, having been found in Norway 180 feet high. The crest of the male flower is larger than in the genus Pinus, as shown at *d* in *fig.* 98, in a magnified side view of one of the cells of a male scale (*a*), from which the pollen has been discharged, the empty case being shown at *c*.  The female scales (*b*) have each

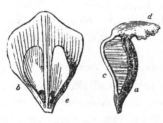

FIG. 98.—THE SPRUCE FIR (*Abies excelsa*).

a small bract at the back, and two seeds inside, (*e*) the wings of which have each a little cavity at the lower part in which the seed lies, so that it is naked on one side, and clothed by the wing on the other. The Spruce Fir bears cones when the trees are of a very small size; and these cones are very ornamental when young, being of a rich purple, while the male catkins are yellow tinged with red at the base. The sap from the Spruce Fir does not flow freely when the bark is wounded, as it does from the Scotch Pine; but oozes out gradually, and is what is called Burgundy pitch in the shops. Spruce-beer is made from the young shoots of the American Black Spruce. There are many species of Abies, but the most interesting are *Abies Douglasii*, a very handsome tree only lately introduced, of very quick growth; and the Hemlock Spruce (*Abies canadensis*). This genus is called Pinus by the Linnean botanists, Picea by Professor Link and some German botanists; and Dr. Lindley, who calls it Abies, includes in it the Silver Fir, the Larch, and the Cedar.

### THE GENUS PICEA.—THE SILVER FIR.

This genus, which includes all the Silver Firs, is readily distinguished from Abies by its leaves, which grow in two rows, one on each side the branch; thus the branch lies quite flat when placed on a table, or any other level surface. The leaves are also not the same on both sides

as in the Spruce Firs, but the under side is
marked by two distinct lines of silvery white,
one on each side the midrib.  The cones stand
erect, and the dorsal bract is frequently so large
as to appear above the tips of the scales ; and
the scales themselves are deciduous, falling off
when the seeds are ripe, leaving the central
rachis bare.    This last peculiarity is very
striking; as both the Pines and Spruce Firs retain
the scales of their cones after the seeds have
fallen.   The seeds of the Silver Firs are much
larger than those of the Spruces ;  and they are
not attached to the wing in the same manner.
The Silver Fir is a noble tree, and takes its
name from the epidermis of its bark; which in
young trees is of a whitish grey, and smooth,
though when the tree is about fifty or sixty
years old, it cracks and peels off, leaving the
dark brown rough bark beneath.   The cones
are produced two or more together ; they are
upright and cylindrical, being nearly as large at
both ends as in the middle.   The leaves all
curve upwards at the point, thus showing con-
spicuously the white lines on the under side.
A remarkable circumstance connected with this
tree is, that when it is cut down, the stump will
remain alive for many years, and even increase
considerably in size, without producing any
leaves or branches.   One in the forests of the

Jura, which was ascertained to have lived ninety-two years after the tree had been cut down, had completely covered the section of the wood with bark. Strasburgh turpentine is produced from this tree. There are several species of this genus, some of which, as for example, *Picea Webbiana*, do not show the dorsal bract; while others, as *P. nobilis*, and *P. bracteata*, have it so large as to make the cone appear quite shaggy. All the species abound in resin, which frequently exudes from the cones. This genus is called Abies by Professor Link, and the German botanists.

### THE GENUS LARIX.——THE LARCH.

This genus consists of only three species, which are easily distinguished from the other Abietineæ by their losing their leaves every winter. The common Larch (*Larix europæa*) is a very handsome tree, with drooping branches, and foliage of a yellowish green, which dies off of a red tinge in autumn. The leaves are linear, and they are produced in tufts in a sort of woody sheath, some of them appearing in the same sheath with the female catkins. The male catkins are smaller, but appear in the same manner. The cones are small, and show the dorsal bracts when young, but when ripe they are seldom visible. The seeds are winged,

and so very small, that it appears wonderful
that a tree frequently above a hundred feet high
can spring from them. The cones are of a
bright red when young, but they become brown
when ripe. The Larch grows very rapidly, and
in situations where no other tree would thrive.
Its wood is very valuable, and its bark is nearly
as useful for tanning as that of the oak. The
trees, however, in some situations are subject to
a disease called pumping, by which the centre
of the trunk becomes as hollow as though it
were intended for a pump. The sap of the
Larch produces the Venice turpentine ; and in
some parts of France a kind of gum, called the
Manne de Briançon, which is used medicinally,
is collected from the leaves.

### THE GENUS CEDRUS.——THE CEDAR.

THERE are only two species in this genus, the
Cedar of Lebanon (*Cedrus Libani*), and the
Deodar (*C. Deodara*). The male catkins of the
Cedar of Lebanon are produced singly, and
each scale has a large crest. The cones are
ovate, and the scales, which are very short and
broad, fall with the seeds, in the same manner
as those of the Silver Fir. The leaves resemble
those of the Larch, but they are not deciduous.
The male and female catkins are very often on
different plants ; and the trees attain a con-

siderable age before they produce perfect seeds.
The Cedar is remarkable for the enormous size
of its branches, and for the shelf-like character
they assume. The tree in a living state lasts
several centuries, but the wood is of a very
coarse grain and not at all durable ; and though
the resin appears so abundant in the cones as to
ooze through the scales, there is so little in the
trunk that it is never used for turpentine.

The Deodar Cedar (*C. Deodara*) closely resembles the common Cedar in its catkins and
cones, but the foliage is of a beautiful glaucous
green, and the leaves are so much longer as to
give a peculiarly graceful character to the tree.
The wood is remarkably durable, very fragrant,
and of an extremely fine grain, taking so bright
a polish, that a table which Mr. Lambert had
of it in his drawing-room has been compared to
a slab of brown agate. The trunk abounds in
resin, and it produces in India a great quantity
of fluid turpentine, which though it is of rather
a coarse quality, is much used by the natives;
pitch and tar are also produced by charring
the wood. The tree on the Himalayas grows
above 150 feet high, with a trunk 30 feet or
more in circumference, and it is said to live to
a great age. It was only introduced into
Britain in 1822, but there are numerous specimens of it in different parts of the kingdom, all
of which appear quite hardy.

### THE GENUS ARAUCARIA.

*Araucaria imbricata*, the only hardy species, is a very singular tree. The trunk is quite straight, with a strong leading shoot, and whorls of branches of great length, and far apart from each other, covered closely with scale-like leaves. These large horizontal arms, clothed with closely imbricated leaves, resemble, in the young trees, snakes partly coiled round the trunk, and stretching out their long, slender, flexible bodies in quest of prey. The male and female flowers are on different trees. The male catkins are cone-shaped, the scales serving as filaments to the anthers produced at their base. The cone is round and very large, with numerous wedge-shaped scales, and large eatable seeds or nuts, which have each a short, callous, marginal wing. The trunk is covered with a very thick corky bark ; the wood is white, finely grained, and durable. The trees when wounded yield a milky juice, which hardens into a fine yellow resin ; and the kernel of the nut, which is as large as an almond, is used by the Indians as an important article of food. The tree is a native of the Andes of Peru, and when first introduced it was called the Chilian Pine. It has now become quite common in this country, and the Earl of Harrington has planted an avenue with it at Elvaston Castle.

There are several species, but the other kinds are too tender to bear British winters without protection. The Norfolk Island Pine (*A. excelsa*) is a splendid tree, with light feathery foliage; as is the Moreton Bay Pine (*A. Cunninghami*).

---

### § II. CUPRESSINEÆ.—THE CYPRESS TRIBE.

MOST of the plants contained in this section are evergreen shrubs or low trees, but some of them attain a considerable size. Only one species, the deciduous Cypress, loses its leaves in winter. Many of the species are only half-hardy in Britain, and none of them are grown in this country for their timber. They all exude resin occasionally from their leaves and branches, but none of them produce turpentine. The catkins are but few flowered, and the cones are roundish. The leaves are frequently imbricated, at least when young; though in many of the species they vary considerably, even on the same tree.

### THE GENUS THUJA.—THE ARBOR VITÆ.

THERE are several species of this genus, but only two are common in British gardens. Of these the American Arbor Vitæ (*Thuja occidentalis*) is the largest tree; though it seldom

grows above 30 feet high, and it is a great
many years before it even attains that height.

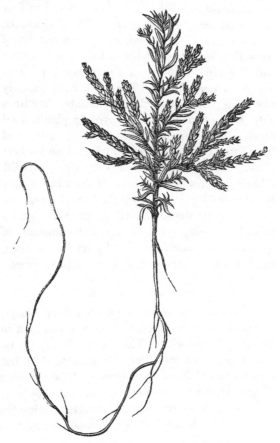

Fig. 99.—Young plant of the Arbor Vitæ.

The male flowers and the female flowers are distinct, but on the same tree. The male catkins are small cones, with the pollen inclosed in four cases that are attached to the inside of the scale, near its base. The female catkins consist of six scales, with two ovules at the base of each; and the ripe cone has a sharp point projecting from each scale. The seeds have scarcely any wing; and when they germinate, they have only two cotyledons. The young plants send down a very long tap root (see *fig*. 99), and have some of their leaves imbricated and others loose. The Chinese Arbor Vitæ (*T. orientalis*) seldom reaches the height of 20 feet, but it may be also distinguished from the preceding species by its more dense habit of growth, by its branches being turned upwards instead of spreading horizontally, and by its leaves being smaller, closer together, and of a lighter green.

## THE GENUS CALLITRIS.

CALLITRIS is a genus separated from Thuja, of which only one species is as yet common in British shrubberies. This is the Gum Sandarach-tree, formerly called *Thuja articulata*, but now named *Callitris quadrivalvis*. The branches of this tree are articulated, that is, they may be broken off at the joints without lacerating the bark. The leaves are very small, quite flat, and articulated like the branches. The male catkins form a cone, in which the scales are disposed in four rows, with three or four anthers at the base of each. The female catkins are solitary, and they divide, when ripe, into four woody valves or scales, only two of which bear seeds. The seeds are small, and have a wing on one side. The tree is a native of Morocco and Barbary, in which countries it produces the gum-sandarach, which exudes like tears from every part of the plant. The wood is fragrant, very finely grained, and extremely durable, as is shown in the roof of the Cathedral of Cordova, built in the ninth century, which is of the wood of this tree.

### THE GENUS CUPRESSUS.—THE CYPRESS.

THE evergreen cypress (*Cupressus sempervirens*) is a cone-like, tapering tree, with its branches growing close to its trunk, and rarely attaining the height of fifty feet even in its native country. The male catkins are longer than those of the arbor vitæ, and the female ones contain more ovules. The cone is buckler-shaped, and it divides, when ripe, into eight or ten corky scales, each of which has four nuts attached ; the cone being partially divided into cells, which may be seen, when the scales have been removed to show the interior. The pollen of each male

flower is contained in four cells, attached to the
lower part of the inside of the scales.    The
wood is remarkably hard and fragrant, and it is
of a fine close grain ; it is also very durable.
It is supposed to have been the gopher-wood of
Holy Writ, and the citron-wood of the ancient
Romans, the beauty of which in tables was so
celebrated.

The White Cedar (*Cupressus thyoides*) is a species of Cypress, having imbricated leaves, and
the same kind of cone ; and the Cedar of Goa
(*C. lusitanica*) is another species of Cupressus,
which appears from the shape of its cones to be
nearly allied to the Arbor vitæ.    There are some
other species, but they are not common in
British gardens.

THE GENUS TAXODIUM.——THE DECIDUOUS CYPRESS.

The Deciduous Cypress (*Taxodium distichum*)
has numerous leaves arranged in two even rows,
one on each side the branch, which fall off in
autumn, assuming a reddish tinge before they
drop.    This genus was separated from Cupressus,
because the male catkins, instead of being produced singly at the tips of the branches, are in
clusters or panicles, and the anther-like scales,
have the pollen in five cells.    The cone, which
is very small, has only two seeds to each scale,
instead of four ; and the young plant has five

or more cotyledons, while the Cypress has only
two.   The deciduous Cypress was placed in the
genus Cupressus by Linnæus, and afterwards it
was called *Schubertia disticha* by Mirbel.   The
tree, which grows 120 feet high and upwards in
America, with a trunk forty feet in circumference
at the base, has generally, when of this size, the
lower part of its trunk hollow, often to the
height of five feet or six feet from the ground.
The roots also send up conical protuberances
two feet high, and four feet or five feet wide,
which are always hollow.   These curious knobs
are called in America " cypress knees ; " and
the negroes use them for bee-hives.   The wood
of the deciduous Cypress is used in building in
Virginia.   There is another species (*T. sempervirens*) which does not lose its leaves in winter,
a native of California, but it has not yet been
introduced.

### THE GENUS JUNIPERUS.—THE JUNIPER.

The species of this genus are extremely variable in their leaves, which differ exceedingly on
the same plant, and in the size to which the
plants attain ; as even the common Juniper,
though generally a shrub not above three feet
high, sometimes becomes a tree.   In the common Juniper (*Juniperus communis*) the leaves
are narrow and pointed, and they are placed in

Q

whorls, three in each, round the branches. The
male and female flowers are generally on different
plants, but sometimes on the same. The male
catkins are sometimes at the end of the shoots,
but generally they spring from the axils of the
leaves. The pollen cases vary from three to
six, and they are attached to the back of each
scale, which may be called the stamen (see *a* in
*fig.* 100). The female catkin, when young,

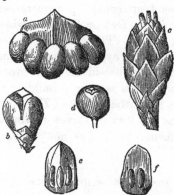

resembles a very
small bud, and
consists of three
fleshy ovaries,
almost hidden
by the thick
scales at their
base. These
ovaries grow
together, and
soon present
the appearance

FIG. 100.—JUNIPER (*Juniperus communis*). shown, but magnified, at *c*. As they ripen, they rise out of the
scales and become the fleshy strobile, *b;* and
finally the spongy berry shown of its natural size
at *d*, containing three seeds or nuts, each of which
is flat on one side, *f*, and angular on the other, *e*,
with five glandular indentations at its base.
The berries are first green, but they afterwards

become of a dark purple, and are covered with a fine bloom. The Juniper berries are very fragrant, and the glands in their stones contain a kind of oil. These berries when crushed are used in making gin and hollands.

There are a great many species of Juniperus, but one of the most remarkable is the Red Cedar (*J. virginiana*). This is a tree forty feet or forty-five feet high. The leaves, when young, are scale-like; but when older they become loose and feathery, so that there are two kinds of leaves on the same tree. The male and female flowers are very small, and the berry is only two-seeded. The sap-wood of this tree is quite white, but the heart-wood is red, and it is used occasionally for making black-lead pencils, particularly those of the commoner kinds, though the Bermuda Juniper is preferred for the superior ones. This last species (*J. bermudiana*) is rather tender in England, and it is seldom grown in this country. Its berries are of a dark red, and they are produced at the ends of the branches; and the wood has so strong a fragrance that shavings of it are put in drawers to keep away moths. The Savin (*J. Sabina*), and several other species, have the old leaves scale-like, as well as those on the young wood. All the species have berry-like fruit, which is generally purple or dark red, and which varies principally in the number

of stones or nuts that it contains.   The fruit of
all the Junipers is very slow in ripening, and in
some of the species it remains two years on the
tree.

### § 3.   TAXINEÆ.—THE YEW TRIBE.

The only needle-leaved tree in this section is the
Yew, and this is the only one I shall describe;
as though the Salisburia and some of the New
Zealand resinous trees are included in it by
modern botanists, the latter are at present very
rare in this country; and the Salisburia, though it
has been introduced more than a hundred years,
and is frequently found in shrubberies, has not
yet produced fruit in Britain.

### THE GENUS TAXUS.—THE YEW.

The common Yew (*Taxus baccata*) has the

FIG. 101.—THE COMMON YEW (*Taxus baccata*).

male and female flowers on different plants.

The catkins of the male flowers consist of a number of scales, out of which the anthers grow like a cluster of primroses, as shown, magnified, in *fig.* 101 at *a*. The female flowers somewhat resemble those of the Juniper ; the ovary being enveloped in scales (*b*), from which it gradually emerges as it swells (*c*), till at last, when ripe (*d*), it opens at the top, and displays the ripe nut enveloped in a red juicy cup. The wood of the Yew is remarkably tough, and the growth of the plant is very slow.

To these may be added the very singular plants comprised in the order Cycadaceæ, which are on the debatable ground between the exogenous and endogenous plants. They bear cones like the pines and firs, but in their leaves, and the manner in which they unroll them, they resemble the ferns, and in the outside of their stems the palms ; while from the wood being in concentric circles, they must be classed among the Exogens. It would be unsuitable to a work like this to enter into any of the discussions of botanists respecting these curious plants ; it may be sufficient here to say that they are considered to be trees, the central cylindrical part being called the trunk ; the soft pith in which, in some of the kinds of Cycas, is manufactured into a spurious sort of sago. The roughness on the stem arises from the remains of the footstalks of old leaves. The

leaves are pinnate, and unroll instead of unfold-
ing.  The flowers are male and female, both
of which are produced in cones in Zamia; and
the male flowers in cones in Cycas, while the
female ones appear on the margin, and in the
notches of abortive leaves, which spring in a
mass from the centre.

# MODERN BOTANY FOR LADIES.

## PART II.

### SYNOPSIS OF THE NATURAL SYSTEM ACCORDING TO PROFESSOR DE CANDOLLE.

### INTRODUCTION.

ALL plants are by this system first divided into the Vasculares and the Cellulares; and to explain the difference between these two great divisions, it will be necessary to say a few words on the construction of plants, though this subject belongs properly to vegetable physiology. All plants are composed of two kinds of matter: viz. Cellular Tissue, which may be compared to the flesh of animals; and Vascular Tissue, which consists of spiral vessels and ducts, which may be compared to the nerves and veins. If any one will take the leaf of a hyacinth and break it by doubling it first on one side and then on the other, and will then draw the parts gently a little way asunder,

the spiral vessels will be seen distinctly with
the naked eye; as though very fine, they are
sufficiently strong to sustain the weight of the
lower part of the leaf for a short time. Now
these vessels are much more conspicuous in
some plants than in others; and in some of the
inferior plants, such as lichens and fungi, they
are wanting altogether. Their presence or
absence has therefore been chosen to mark the
two great divisions of the Natural System;
the Vasculares being those plants which have
both vascular and cellular tissue, and the Cellulares being those which have cellular tissue only.

The Vasculares are much more numerous
than the Cellulares; and they are redivided into
subclasses, which it also requires the aid of vegetable physiology to explain. All plants, when
in a growing state, require the moisture taken
up by their roots to be elaborated, and mixed
with air in the leaves before it becomes sap;
that is, before it is fit to contribute to their
nourishment. Now, when a seed begins to germinate, its root descends into the ground and
its plumule, or ascending shoot, rises upwards;
but the leaves folded up in the latter are too
weak and tender to elaborate the sap; and
besides, they cannot act till they are fully
expanded, and they want nourishment to give
them strength to unfold; the roots are also not

sufficiently developed to absorb moisture. To supply the wants of the young plant while it remains in this feeble and imperfect state, a quantity of albumen is laid up in the seed; and it is evident that this substance must be extremely nourishing, as it forms, when ground, what we call flour. The young plant is thus provided with food, till its roots are sufficiently developed to obtain it from the soil; and to enable it to elaborate this food and to turn it into sap, it is, in most cases furnished with one or more seed-leaves, or cotyledons, (see *c* in *fig.* 102,) which drop off as soon as the true leaves are sufficiently advanced to be able to perform their proper functions. The cotyledons differ in number, form, and substance in the different genera; but in all plants, they are very different from the true

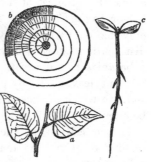

FIG. 102.—COTYLEDONS, LEAVES, AND WOOD OF A DICOTYLEDONOUS PLANT.

leaves, being admirably contrived for answering the end for which they are designed; and it having been found that the plants having two or more cotyledons differ widely in many other respects from those having only one cotyledon, the number of the cotyledons has been chosen as the distinguishing mark of the second great division of plants according to the Natural System.

The Vascular plants are therefore again divided into the Dicotyledonous plants, which have two or more cotyledons; and the Monocotyledonous plants, which have only one cotyledon: and to these modern botanists add the Acotyledonous plants, which have no cotyledon, as some of them have spiral vessels, or at least ducts, though most of the Acotyledonous plants belong to the Cellulares. These divisions are not only marked by their cotyledons, but they are so distinct in other respects, that it is sufficient for a botanist to see a leaf, or even a bit of wood without leaves, of any plant, to know at first sight to which of these three divisions the plant belongs.

If the leaf of a Dicotyledonous plant be examined, it will be found to have a strong vein up the centre, from which other veins proceed on each side (as shown at *a* in *fig.* 102); and if it be held up to the light, the rest of the leaf will be found to be intersected by numerous smaller veins, so as to appear like network, and hence these leaves are said to be reticulated. The trunk and branches of trees belonging to this division consist, when young,

of pith, wood, and bark. At first the substance
within the bark is little more than pith, but as
the returning sap deposits every year a layer of
wood just within the bark, which presses against
the previous layers so as to contract them, they
press in turn against the pith, which becomes
smaller and smaller every year, till at last, in
old trees, it is scarcely perceptible. The layers
of wood are always perfectly distinct from each
other, and they are called concentric rings (*b*) ;
while faint lines, with which they are intersected,
and which proceed like rays from the remains
of the pith in the centre, are called medullary
rays. A layer of wood being deposited every
year, the age of the tree may be discovered by
counting the concentric rings ; also if the tree
has grown rapidly, the layers will be thicker
than if it has grown slowly ; and if it has had
one side more exposed to the sun than the other,
the remains of the pith will be on one side
instead of in the centre, and the layers will be
thinner on that side than on the other. The newest layer of wood, which is called the alburnum
or sap-wood, is of a paler colour and more porous
texture than the rest of the tree, and it is of less
value as timber. It is from the manner in which
the successive layers of wood are deposited that
Dicotyledonous trees are called exogens, which
signifies, to increase from the outside.

If the leaf of a Monocotyledonous plant be
held up to the light, the principal veins will be
found to form parallel lines of nearly equal
thickness, the central one being very little, if any,
larger than the others
(see *a* in *fig*. 103).

The trees belonging to
this division are all
natives of the tropics,
and their softest and
newest wood is in the
centre, where fresh de-
posits are made every
year inside the old
wood; and hence, these
trees are called endo-
gens, which signifies, to
increase from within.
The wood of these trees

FIG. 103.—MONOCOTYLEDONOUS
PLANT.

has neither medullary rays nor concentric rings;
and a section of it appears pierced with numerous holes (*b*), as may be seen by cutting off a
slice of bamboo.   The germination of a Monocotyledonous plant, with the cotyledon remaining in the ground, is shown at *c*.

The Dicotyledones and the Monocotyledones
have all visible flowers, and are hence called
Phanerogamæ; but the Acotyledones have no
visible flowers, and they are hence called Cryp-

togamæ, which signifies that their flowers are hidden. The most remarkable of the Crypto-gamous plants are the ferns, some of which become lofty trees; the wood of which is in curious wavy lines, as it appears to be formed by the footstalks of the decayed leaves growing together and becoming woody. The veins in the leaves or fronds of the ferns are forked.

Besides the great divisions already mentioned, the Dicotyledonous plants have been divided into the Dichlamydeæ, or those having both calyx and a corolla; and the Monochlamydeæ, or those having only a calyx; but there are so many exceptions, as to render this division of little value. The Monochlamydeæ are not sub-divided, but the Dichlamydeæ are again divided into the Thalamifloræ, in which the petals and the stamens grow separately out of the thalamu or flat part of the receptacle, and generally from below the pistil; the Calycifloræ, in which the stamens and petals are either attached to the calyx, or to a lining of it formed by the dilated receptacle; and the Corollifloræ, in which the petals grow together, so as to form a cup for the pistil, and which have the stamens attached to the corolla, but quite distinct from the calyx. The Monocotyledones have also been redivided into the Petaloid, or those with regular flowers, like the bulbous plants, and Orchidaceæ; and the Glumaceous plants, or those with scales or glumes instead of petals, as in the sedges and the grasses. The Acotyledones are divided into those with leaves, as the ferns; and those without leaves, as the mosses, lichens, and fungi.

I have only to add that each subclass is divided into numerous orders, which are differently arranged by different botanists; the object being to place those nearest together which are most alike. As no one of these arrangements appears to be decidedly better than the others, I have adopted that given in Mr. Loudon's *Hortus Britannicus;* marking, where, they occur, the new orders which have been formed, and the alterations in the old ones that have been made since that work was written.

# CHAPTER I.

PHANEROGAMOUS PLANTS—DICOTYLEDONÆ—I. DICHLAMYDEÆ.
§ I.—THALAMIFLORÆ.

In all the plants contained in this chapter the receptacle is a fleshy substance called the thalamus, or disk, which is surrounded by the calyx, and out of which the carpels or seed-vessels, the stamens, and the petals, all grow separately from each other. Sixty-five orders are included in this division, but I shall only describe those which contain plants which have been introduced into Britain, except where the orders chance to contain plants well known in commerce.

## ORDER I.—RANUNCULACEÆ.

The plants belonging to this order are known by their numerous stamens, the anthers of which burst outwardly; by their carpels growing close together without adhering, except in one or two instances; and by the stem-clasping petioles of their leaves, which are generally deeply cut. The flowers when regular have five petals and five sepals, but they differ widely in shape, and the calyx of several of them is coloured so as to resemble a corolla. The seeds are fre-

quently cariopsides ; and the plants abound in
a watery juice which is acrid, and in most cases
poisonous.

## ORDER II.—DILLENIACEÆ.

THIS order resembles Ranunculaceæ in having
five petals, five sepals, and numerous stamens ;
but the anthers burst inwardly instead of outwardly, and there are never more than five
carpels, and seldom more than two, which often
grow into a berry-like fruit, as in the genus
Dillenia from which the order takes its name.
One species of this genus is occasionally seen in
English hothouses, *Dillenia speciosa*. It has
yellow flowers with the five petals apart at the
base, and the sepals edged with white. The
fruit consists of five carpels growing together
with a sort of crown formed by the spreading
stigmas. Another genus, some of the species of
which are found in British greenhouses, is
*Hibbertia*. The species are generally climbing
plants, with flowers like those of Dillenia, but
smaller, though *H. dentata* has the petals close
together. The difference between the genera
consists principally in the carpels, which in
Hibbertia are distinct with long filiform styles
curving inwards. All the plants contained in
this order are evergreen exotic shrubs and trees
with simple alternate leaves, and, with only two
or three exceptions, the flowers are yellow.

ORDER III.—MAGNOLIACEÆ.

THIS order was divided by De Candolle into two tribes: viz. *Illicieæ*, the Aniseed tribe; and *Magnolieæ*, the Magnolia tribe. The first, which is now made a distinct order, under the name of Winteraceæ, contains three genera, only one of which, Illicium, is common in this country. The only hardy species of this genus, *I. floridanum*, the Florida Aniseed tree, has very dark purple flowers, which appear to be double from the great number of the petals, which are from twenty to thirty. The carpels are also numerous, and arranged so as to form a star. All the plants in this tribe are highly aromatic, and one species, *Drimys Winteri*, which has white fragrant flowers, produces an aromatic bark that is used in medicine.

The tribe Magnolieæ is distinguished by the fruit consisting of a number of carpels arranged so as to form a cone. There are six genera in this order, the most remarkable of which are Magnolia, Liriodendron, Talauma, and Michelia, the last genus consisting of stove trees, with very fragrant flowers, which are generally of a pale yellow, and only one species of which, *M. Champaca*, has been introduced.

Of these genera Magnolia is undoubtedly the best known; as nearly all the species are common in British gardens. This genus is divided into two sections, one containing the American Magnolias, and the other those from Asia, which are principally from China and Japan.

The latter may be illustrated by *Magnolia conspicua*, sometimes called *M. Yulan*. The flower-buds are inclosed in a brown hairy case formed of two short bracts which become loose at the base, and are pushed off by the expanding flower. The flower itself (see *fig.* 104) is cup-

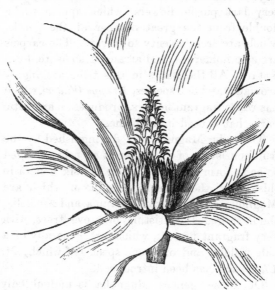

Fig. 104.—FLOWER OF MAGNOLIA CONSPICUA OPENED TO SHOW THE STAMENS AND PISTIL.

shaped, and it is divided into six white fleshy petals. The calyx consists of three sepals, which fall off soon after the petals expand. In the centre of the flower is the receptacle, drawn up into a fleshy cone, with a great number of carpels attached to it, each of which has one cell containing two ovules, and a curved stigma. Around this cone grow the stamens, with very long anthers standing up like palisades, and very short thick filaments. The fruit is oval, with the ovaries somewhat distant from each other. The flowers appear before the leaves. The other Asiatic species are *M. gracilis* or *Kobus*, *M. discolor*, *obovata*, or *purpurea*, and *M. fuscata*; the former two forming handsome shrubs in the open ground, and having cup-shaped flowers which are white within and purplish on the outside, and the latter being a greenhouse plant, with brown very fragrant flowers.

The American species of Magnolia differ in having their flower-

FIG. 105.–THE FLOWER-BUD OF THE EVERGREEN MAGNOLIA.

buds enveloped in one long spathe-like bract,
as shown in *fig.* 105.   The ovaries grow close
together;  and, when ripe, the carpels, which

look like the scales of a
fir-cone (see *fig.* 106),
burst by a slit down the
back;   and   the   seeds,
which are covered with a
red juicy pulp, burst out,
and hang down by a long
white thread, which in
the course of a few days
withers away. The princi-
pal species of American
Magnolias are the ever-
green Magnolia, or Big
Laurel (*M. grandiflora*);
the Umbrella Tree (*M.
tripetala*), which   grows

FIG. 106.—THE RIPE FRUIT AND
SEEDS OF THE EVERGREEN
MAGNOLIA.

like a shrub with several stems rising from the
ground ; the Cucumber-tree (*M. acuminata*), the
flowers of which are bluish and the leaves pointed ;
Beaver wood (*M. glauca*), the flowers of which
are small, and the leaves covered with a glaucous
bloom ;  *M. auriculata, M. pyramidata* and *M.
macrophylla*, which are nearly allied to the Cucumber-tree; and *M. cordata*, the flowers of which
are yellowish.  All these Magnolias produce their
leaves before their flowers; and in this also they

differ from *M. conspicua*, the flowers of which appear before the leaves.

The genus Liriodendron contains only two species differing slightly in the leaves. Both are lofty trees, with cup-shaped flowers of six petals curiously stained with red and yellow, and bent back at the tip. The calyx consists of three sepals, which remain on as long as the petals. The fruit is cone-shaped, but the carpels, which are each furnished with a kind of wing, instead of opening when ripe, fall with the seed enclosed.

The genus Talauma differs from Magnolia principally in the carpels, which open irregularly by valves; and in the number of petals, which vary from six to twelve. Only two species are common in British hothouses, *T. Candolli*, commonly called *Magnolia odoratissima;* and *T. pumila*, sometimes called *M. pumila* and sometimes *Liriodendron lilifera :* both are natives of Java, and both have cream-coloured, or yellowish flowers, which are remarkably fragrant at night.

ORDER IV.   ANONACEÆ.—THE CUSTARD-APPLE TRIBE.

THE hardy plants belonging to this order, that are well known in Britain, were formerly included in the genus Anona; but now the only species retained in that genus are stove plants, natives of the West Indies, with yellowish brown or dark purple flowers, the calyx of which is in three sepals, and the corolla in three or six thick fleshy petals, and which have numerous stamens with large angular anthers, and very short filaments. The carpels are numerous, but they grow altogether into a fleshy eatable fruit, divided into many cells, each containing one seed. This fruit is called the custard apple or sour sop in the West Indies; and it differs in flavour in the different species, but the most delicious kind is produced by *A. Cherimolia*, a native of Peru. The hardy species included in Anona by Linnæus have been separated from that genus, and formed into another under the name of *Asimina*, the principal distinction between them being in the fruit; which in the genus Asimina consists of two or three berry-like carpels growing together, not eatable, and each containing many seeds. *A. triloba*, the hardiest species, is a large shrub, with dark brownish purple flowers. The plants in this order are all aromatic.

---

### ORDER V.—MENISPERMACEÆ.—THE COCCULUS TRIBE.

ALL the plants contained in this order are climbing exotic shrubs, generally with drooping racemes of small delicate flowers, the male and female flowers being on different plants. The number of sepals and petals varies in the different genera, and sometimes the petals are wanting. The stamens frequently grow together into a central column; and the fruit is a drupe or one-seeded berry, generally scarlet, but sometimes black. The principal plants in this order which are known in England, are, *Menispermum canadensis* (the Canadian Moon Seed), a very ornamental hardy, climbing, shrub; *Cocculus palmatus,* the root of which is a tonic drug, called Columba root; *Anamirta Cocculus,* which produces the berries called *Cocculus indicus* in the shops, which are said to be used in porter to give it an intoxicating property; *Schizandra coccinea,* a greenhouse climber with scarlet flowers; and *Kadsura japonica,* a climbing shrub with white flowers and red berries, which proves quite hardy in the open air. *Kadsura, Schizandra,* and three other genera, little known in this country, have been formed into a new order under the name of Schizandriaceæ. The qualities of all these plants are tonic.

---

ORDER VI. BERBERIDEÆ.—THE BERBERRY TRIBE.

EACH flower of the common Berberry (*Berberis vulgaris*) has on the outside three little bracteal scales, which are reddish on the back, and soon

fall off. The flower itself consists of a corolla
of six petals, and a calyx of six sepals, though
as these divisions are all of the same size and
shape, and of the same colour and texture, it is
not very easy to distinguish the calyx from the
corolla. The petals will however be found on

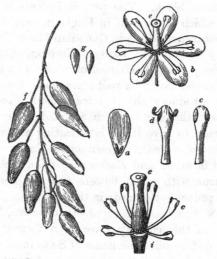

Fig. 107.—Details of the flower and fruit of the common
Berberry, partly magnified.

examination to have each two little glands at
the base, as shown at *a* in *fig.* 107, which the
sepals are without. The sepals are placed exactly behind the petals, so that the one appears
a lining of the other ; and, being concave, the

petals serve as a kind of cradle to the stamens, as shown at *b*. There are six stamens, which have broad filaments; and instead of anthers the filaments are widened at the tip, and each contains two cases for the pollen (*c*); these cases are each furnished with a valve-like lid (*d*), which opens and curls back when the pollen is ripe. The pistil (*e*) is pitcher-shaped, with a very thick style, and a flat stigma. It stands erect, while the stamens are spread out so as to be a long way from it, but they are so irritable that the slightest touch makes them spring forward and discharge their pollen on the stigma, afterwards falling back into their former places. The flowers are yellow, and they are produced in long drooping racemes; and they are succeeded by red oblong berries (*f*), each of which contains two seeds (*g*). The receptacle, with the stamens growing out of it from beneath the pistil, is shown at *i*. The common Berberries are all deciduous shrubs, with simple leaves, which are produced in tufts, as shown in *fig.* 108, each leaf being delicately fringed with hairlike teeth. Each tuft of leaves has two or three sharply-pointed stipules, which are easily distinguished from the leaves, by their margins being without teeth; and below these are three spines, which, when young, are soft and look like folded leaves, but which, when older, become hard, and sharply pointed. These spines are considered by some botanists to be abortive branches. There are many different kinds of

FIG. 108.—FLOWERS OF THE BERBERRY, NATURAL SIZE.

Berberry, which differ principally in the size of the flowers and in the colour of the fruit; but which also vary in the size and shape of the leaves, and in the manner in which they are toothed.

The Ash-leaved Berberries were formed into a separate genus called Mahonia by Nuttall; and this genus has been adopted by Professor de Candolle, and other botanists. Dr. Lindley, however, includes all the species in the genus Berberis, and he has been followed by Mr. George Don in his new edition of Sweet's *Hortus Britannicus.* Whether the genus Mahonia be a good one or not, the plants composing it are very distinct from the true Berberries. The leaves of the Mahonias are evergreen, and pinnate; and the leaflets instead of being fringed with fine hairs, are broadly serrated, the points being tipped by a sharp prickle or mucro (see *a* in *fig.* 109); and the petiole is articulated, and somewhat stem-clasping at the base (*b*). The flowers are in erect racemes, and smaller than those of the Berberry; they are also more globular, being less widely opened, and the petals are without any glands. The filaments of the stamens have two hair-like teeth just below the lobes of the anthers; and the fruit has from three to nine seeds in each berry; while the Berberries have only one or two. There are many kinds of Magnolia, but the handsomest is *M. Aquifolium*,

FIG. 109.—A COMPOUND LEAF, BEING THAT OF MAHONIA AQUIFOLIUM.

a hardy shrub, with dark green shining leaves, like the holly. All the species both of Berberis and Mahonia have yellow flowers; and the Mahonias all flower very abundantly, and very early in spring.

The principal other plants belonging to this order are, *Nandina domestica*, a very pretty shrub with white flowers, from China, which requires a greenhouse in England; several species of Epimedium, some of which are from Japan, with purple and white flowers; a few species of Leontice, pretty plants with yellow flowers; and a plant called *Diphylleia cymosa*, with white flowers and blue berries, a native of North America. All these plants are easily re-cognised by their broad stamens, and the curl-ing back of the valves of their anthers.

## ORDER VII. PODOPHYLLACEÆ.—THE MAY-APPLE TRIBE.

THIS order contains only two genera; viz., Podophyllum and Jeffersonia; both of which have a calyx of three or four sepals, and a white corolla of from six to nine petals. Podophyllum has numerous stamens, and a fleshy berry with only one cell, which does not open when ripe; and Jeffersonia has eight or nine stamens, and a capsule which opens all round the apex. *Podophyllum peltatum* is the May-apple, and its fruit is eatable when ripe, though very acid; the

leaves are very large, and peltate, that is, with
the footstalk attached to the centre; and
*Jeffersonia diphylla* is a little plant, without any
stem but that which supports the flower. Both
are natives of America, where they are found
in moist shady places.

---

### ORDER VIII. HYDROPELTIDEÆ.

THIS order, which many botanists combine
with the preceding one, also consists of only
two genera; viz., *Cabomba* and *Hydropeltis*;
and of these *Cabomba aquatica* is a stove aquatic,
and *Hydropeltis purpurea* is a hardy water plant,
with peltate leaves, and dull purple flowers.

---

### ORDER IX. NYMPHÆACEÆ.—THE WATER-LILY TRIBE.

THE principal genera in this order are
Nymphæa, Euryale, Victoria, Nuphar, and
Nelumbium. The flowers of the common White
Water-lily (*Nymphæa alba*) consist of numerous sepals, petals, and stamens, all of which
might be mistaken for petals, being principally
distinguished by their colour. The sepals, (*a*
in *fig.* 110,) are green on the outside, but they
are white within, and of the same fleshy substance as the petals (*b*). The stamens (*c*) look
like narrow yellow petals; they are pointed,

and bear the pollen in two lobes near the point, which open longitudinally when ripe. The inner row of stamens are without anthers, and form a kind of vandyke edging to the pistil, as shown at *e*. The pistil consists generally of sixteen carpels, growing together into a vase-like, many-celled berry, as shown at *d;* the spreading stigmas, which have also grown together, forming a kind of lid. The carpels are com-

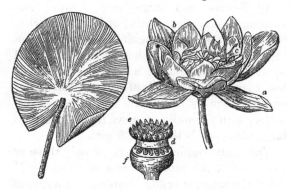

FIG. 110.—THE FLOWER, LEAF, AND SEED-VESSEL OF THE WHITE
WATER-LILY, GREATLY REDUCED IN SIZE.

pletely enclosed by the receptacle which rises up round them, and forms a thick fleshy covering, as shown at *f.* The seeds are numerous, and they are covered with a thick leathery skin. The embryo is small, and it is surrounded by a great mass of floury albumen. The leaves (*g*) are large and nearly round ; and the main root, which is called a rhizoma, is thick and fleshy, and is, indeed, an under-ground stem. There are several kinds of Nymphæa, the most remarkable of which is the Egyptian Lotos (*N. Lotos*), the flowers of which are white tinged with pink ; and both the roots and seeds of which are eaten. *Euryale* is a genus of South American Water-lilies, generally with small flowers, and large rough leaves ; and *Victoria regina*, also a native of South America, is perhaps the most magnificent Water-lily in the world ; the leaf, which is peltate and turned up at the brim, being of a deep crimson on its lower surface, is upwards of six feet in diameter; and the flowers are more than a foot in diameter, with a corolla of more than a hundred large white petals tinged with pink.

The genus Nuphar consists of only three or four species, the most common of which is *N. lutea*, the common yellow Water-lily, a native of Britain. The flower has a cup-shaped calyx of five large yellow sepals, the tips of which curve inwards. The petals are small, truncate, and flat, with a small pore on the back of each ; and the stamens, which are very numerous, have broad petal-like filaments. They differ, however, very much in appearance from those of the genus Nymphæa, and they are differently

placed, springing from the base of the vase-like
pistil, and not from the upper part. There are from
sixteen to twenty carpels enclosed in the dilated
receptacle, to which the stigmas form a ray-
like cover; and each carpel contains several
seeds. The leaves are somewhat cordate, and
rise rather above the surface of the water, and
the rhizoma, or root-stem, is very thick. The
common yellow Water-lily, or Brandy-bottle,
as it is sometimes called from the smell of its
flowers, is common in every part of England,
and it is generally found in small ponds or
ditches. The other species are mostly natives
of North America.

The Indian Lotos (*Nelumbium speciosum*)
differs so much from both the preceding genera,
as to be considered by some botanists to form
a different natural order. The sepals and petals
are so intermingled in the flower as to be
scarcely distinguishable; but the filaments of
the stamens are less broad and petal-like. The
disk is still elevated, but it has lost the vase-like
form, and it appears as though the top had been
abruptly cut off; while the carpels are no longer
joined together, but are plunged each separately
in the fleshy receptacle, or torus, with their
stigmas quite distinct. As the carpels are only
half immersed in the torus, and thus show their
styles and stigmas, they have a very singular

s

and bottle-like appearance; and the torus,
when they are taken out of it, looks like a
piece of honey-comb.  The rhizoma is white
and fleshy.  The stalks of the flowers and leaves
rise considerably above the water; and thus
the flowers have not the graceful appearance of
those of the Nymphæa, which seem to repose
on the surface.  The leaf is very large, being
sometimes one or two feet in diameter; and it
is always peltate, with the stalk exactly in
the centre.  There is only one seed in each
carpel of the Nelumbium; and this seed, which
has no thick leathery skin, is of about the size
and shape of an acorn.  It is very good to eat,
having a sweet milky flavour, and in botanical
construction it resembles the common bean,
having no albumen, but a very large embryo.
This is probably the reason why it has been
supposed to be the bean of Pythagoras, and why
it is called the Sacred Bean of India.  One of
the Hindoo fables represents the god Bramah
as first appearing in the form of a child, cradled
on a Lotos leaf, and floating on the waste of
waters.

There are several kinds of Nelumbium, one
of which, a native of America, has double
yellow flowers; and they all require a stove in
England.

## ORDER X. SARRACENIEÆ.—THE SIDE-SADDLE PLANT.

THERE is only one genus in this order, which can never be mistaken for any other, from the pitcher-shaped petioles of its leaves, and its singular flowers. It is a native of Canada, but it rarely flowers without a stove in England. It is a dwarf plant, and it is thus easily distinguished from the Chinese Pitcher plant, which grows eight or ten feet high, and which belongs to quite a different order.

---

## ORDER XI. PAPAVERACEÆ.—THE POPPY TRIBE.

THIS tribe contains several genera, all of which have a thick glutinous juice when broken, which poisons by stupifying. The genera most common in British gardens are Papaver, the Poppy; Argemone, the Prickly Poppy; Meconopsis, the Welsh Poppy; Sanguinaria, Blood-root; Eschscholtzia; Hunnemania; Rœmeria; Glaucium, Horned Poppy; Chelidonium, Greater Celandine or Swallow-wort; Hypecoum; Platystemon, and Platystigma. Most of these plants are either annual, or last only two or three years, and they have all very handsome flowers, which are generally large and of showy colours.

s 2

The common Corn Poppy (*Papaver Rhœas*) has a showy flower, the corolla of which consists of

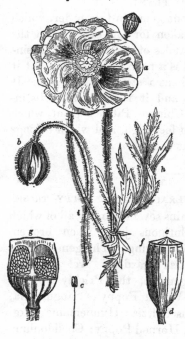

four very large scarlet petals, the outer two much exceeding the others in size (see *a* in *fig.* 111). The calyx is green, and it is divided into only two sepals, (see *b*,) which fall off soon after the expansion of the flower. The petals are all curiously crumpled in the bud, and they present quite a wrinkled appearance when the flowers are first opened.

FIG. 111.—FLOWER, LEAF, AND SEED-VESSEL OF THE COMMON POPPY.

The stamens are very numerous, and the anthers, which are black, are of the kind called *innate ;* that is, the filament is only attached to them at the lower part (*c*). The seed-vessel of the Corn

Poppy is, when ripe, a dry leathery capsule (d)
with numerous angles, each angle indicating a
carpel; for the capsule of the Poppy, though
one-celled when ripe, consists, in fact, of a
number of carpels grown together. The remains of these imperfect carpels are perceptible
in the little valves shown at f, which open at the
top of each to discharge the seed when it is ripe;
and in the slightly-peaked cover (e), which consists of as many stigmas grown together as there
appear to have been carpels. When the capsule is
cut open (as shown in the capsule at g, from
which the fourth part has been removed), remains of the carpels will be found in several projections from the sides, which partially divide
the inside of the capsule into several imperfect
cells, in which the young seeds are formed;
though none of these portions reach the centre.
The ovules, when first formed in the ovary, are
attached to these projections, which are called
parietal placentæ; but as the seeds ripen they
become loose, and if a dry Poppy-head be
shaken, they will be found to rattle. The leaves
of the Corn Poppy are what is called pinnatifid,
(see h in fig. 111,) that is, they are so deeply
cut as to appear almost in separate leaflets;
and the whole plant (except the petals and the
capsule) is covered with short bristly hairs (i),
which stand out horizontally.

The Opium Poppy (*Papaver somniferum*) differs from the Corn Poppy in several respects. First, the whole plant is glabrous, that is, devoid of either hairs or bristles; the capsule also is much larger and more fleshy in an unripe state, and the crown-like lid is smoother, and curved over like a plume of feathers, (see *a* in *fig.* 112.)

FIG. 112.—THE FLOWER, BUD, LEAVES, AND CAPSULE OF THE OPIUM POPPY, MUCH SMALLER THAN THE NATURAL SIZE.

The fleshiness of the unripe seed-vessel of the Poppy puzzled me extremely at first, as I knew that the ripe capsule of this Poppy is always dry and leathery; but it was soon explained to me, that this fleshy substance is, in fact, an elongation of the receptacle or disk, which rises up round the carpels, and envelops them, in the same way as the disk of the Water-lily grows round the pistil and carpels of that plant,

but which dries up as the fruit ripens. The
leaves of the Opium Poppy (*b*) differ from those
of the Corn Poppy, in being much broader, and
only slightly cut or notched; they are also
glaucous, that is, of a bluish or sea-green, and
they are clasped round the stem at their base
(as shown at *c*). All the Papaveraceæ abound
in a thick glutinous juice, which in the Poppies
has the colour and appearance of milk, and
which possesses stupifying properties; but in the
Opium Poppy this juice is particularly abundant.
Opium is, in fact, procured by wounding the
fleshy capsule with a sharp knife, and suffering
the milky juice which exudes to dry in the sun;
after which it is scraped off with a blunt instru-
ment, and pressed into cakes for sale. The
opium of commerce is produced in hot countries;
but even in England, any one who chooses may
procure a small quantity of opium, by wounding
the fleshy capsule of the common White Poppy
when it is about half ripe. A milky juice will
issue from the wound, which when dry becomes
opium, and would be poisonous if taken in
excess. The capsule of the White or Opium
Poppy has, when ripe, a little window-like opening under each stigma for the discharge of
the seeds, which contain abundance of oil, and
may be safely eaten, though the rest of the
plant is poisonous.

There are many different kinds of Poppy ; but
they all agree in the corolla of their flowers
being in four petals, or in some number divisible
by four; and in the calyx, which is generally in
two sepals, dropping off as soon as the flowers
expand.    All the species abound in a milky
juice, which poisons by stupifying ; and they all
agree in the general construction of the capsule,
with its fleshy envelop and its stigma-formed
lid.    The petals are always crumpled in the bud,
and they fall very soon, so that the beauty of the
flowers is very short-lived.    The flower-buds
droop ; but when the flowers expand, the stalk
becomes erect, and remains so while the capsule
containing the seeds is ripening ; a wise provision, common to many plants, to prevent the
seeds from falling too soon.    The calyx of most
of the Poppies is in only two sepals; but in the
two showy perennial species, called *P. orientale*
and *P. bracteatum*, the calyx is in three sepals.

Among the other plants belonging to the
order Papaveraceæ, may be mentioned the
Horned Poppy (*Glaucium luteum*), which,
instead of an obovate capsule, has a long horn-
like pod, divided into two cells, the valves
opening from the top to the bottom.    The
whole plant is glaucous ; and the leaves, which
are broad and notched, clasp the stem at their
base, like those of the Opium Poppy.    The

Prickly Poppy (*Argemone mexicana*) has the whole plant covered with strong prickles ; the leaves are wrinkled and curved up at the margin; the calyx has three sepals; and the capsules are in four or five valves, the stigmas forming a kind of cross at the top.  The stem and leaves when bruised give out a thick glutinous juice, which, instead of being white like that of the Poppy, is yellow.

The Eschscholtzia is the last genus of the order Papaveraceæ that I shall mention here, and it deserves a particular description, both from its popularity and the beauty of its flowers, and from the singularity of its botanical construction.  The bud when it first appears is enfolded in a calyx, which is pointed at its upper extremity, and appears to have a kind of rim near its base.  When the flower is ready to expand, the calyx detaches itself all round from the projecting rim, and rises gradually without opening, till the flower actually pushes it off.  The detached calyx resembles an extinguisher, and hence it is called calyptrate, which has that signification.  The flower is cup-shaped ; there are four petals and four stigmas, two of which are much longer than the others.  The capsules are elongated like those of the Horned Poppy, but they are distinguished by the projection of the flat fleshy disk at their

base; they are two-valved and two-celled.
The leaves are glaucous, and finely cut.  There
are three species, or perhaps varieties, which
differ principally in the degree of enlargement
of the receptacle or disk.  They have all large
fleshy roots, which bleed copiously if wounded,
and for this reason the plants are difficult to
remove unless when quite young.

---

### ORDER XII. FUMARIACEÆ.—THE FUMITORY TRIBE.

THE flowers of plants of this order are so
peculiar in their shape, as when once seen to be
easily remembered.  There are two small
sepals, which soon fall off, and four petals of an
irregular shape, two of them being drawn out
into a kind of spur.  There are six stamens, and
the fruit is silique-formed.  The plants have
somewhat of a smoky smell, and when broken
yield a watery juice.  The principal genera are
Fumaria, Corydalis, and Diclytra.

---

### ORDER XIII. CRUCIFERÆ.—CRUCIFEROUS PLANTS.

THE Cruciferous plants form so natural an
order, that when one of them has been described the others may be easily recognised.
They have all a separate calyx and corolla, each

in four divisions ; the four sepals being placed

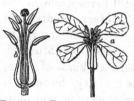

alternately with the four petals, the latter forming a cross (as shown at *a* in *fig.* 113), whence the name of Cruciferous, which signifies cross-

FIG. 113. — THE FLOWER AND
STAMENS AND PISTIL OF THE SEA-
KALE, TO ILLUSTRATE THE CRU-
CIFERÆ.

bearing. There are six stamens (*b*), two of which are much shorter than the others ; and two carpels with one style, and a capitate or divided stigma. The seed-vessel is a kind of pod, either short and broad, like that of the Shepherd's Purse (*fig.* 114), where it is called a sillicle ; or long and narrow, like that of the Cabbage, which is called a silique. The two valves of the silique open naturally when ripe, from the bottom curving upwards, (see *fig.* 115,) and the seeds are deposited on a thin membrane between

FIG. 114.—A
SILLICLE.

the cells, which is the dissepiment. All the Cruciferæ, from abounding in nitrogen, have an unpleasant smell when decaying, like putrid flesh ; and when cultivated, they even in a wild state require abundance of animal manure ; hence, they are generally found near human habitations, or where cattle are kept. When wild, they have generally acrid properties ; and

though these are in most cases softened by
cultivation, yet they are still perceptible in the
roots of the Horse-Radish, and
the common Radish, and in the
leaves and seeds of Mustard,
and the different kinds of Cress,
&c. This acridity, however, is
never so great as to be injurious;
and Cruciferous plants, particularly if their texture be succulent and watery, may always

FIG. 115.—A SILIQUE. be eaten with perfect safety.
Even those which, in a wild state, are
tough and stringy, such as the wild Cabbage
and the root of the wild Turnip, become excellent by cultivation; and all Cruciferous plants
are so extremely nourishing as to be considered
next in this quality to animal food.

Among the many garden flowers which
belong to this order, few are more popular
than the common Wallflower. (*Cheiranthus
Cheiri.*) Its hardiness, and the facility with
which it is raised and cultivated—the gaiety of
its flowers, their profusion, and their delightful
fragrance, combine to make it a general favourite;
and I think I cannot take a flower to illustrate
the order which is more generally known and
liked. The flowers of the Wallflower (see
*fig.* 116, *a*) consist of four petals, each of which

is furnished with a long tapering point, called
the claw (*b*), and a broad flat part called the
limb (*c*).   The claws of the petals are buried in

FIG. 116.—FLOWERS OF THE WALLFLOWER.

a calyx of four sepals, placed alternately to the
petals, and somewhat swelled out at the base,
(see *d*).   The stigma (*e*) is two-lobed, and forms
a kind of notched head.  There are six stamens,
which appear at first to be all nearly of the same
height, but on examination it will be found that
two are somewhat shorter than the others.   The
seed-vessel is of course the lower part of the
pistil; which, after the petals drop, becomes

elongated into a somewhat cylindrical silique,
which contains several flattish seeds.

The Brompton Stock (*Mathiola incana*), and
the Ten-week Stock (*M. annua*), differ from the

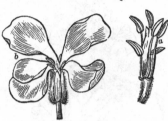 Wallflower prin-
cipally in the
shape of the
stigma (see *fig.*
117, *a*). The
petals have also
rather longer
claws, and hang
looser, as shown

FIG. 117.—FLOWER OF THE BROMPTON STOCK.

at *b*. The Virginian Stock (*Malcomia mari-
tima*) has a roundish silique, and only one
stigma, which ends in a long tapering point.
The Candytuft (*Iberis umbellata*) has a short
pod or sillicle, which has the appearance of
being cut off at the point, and which contains
only one seed in each cell ; and the outer two
petals of the flower are somewhat larger than
the inner ones. Many other Cruciferous plants
might be described, but I think my readers will
have pleasure in seeking them out themselves,
and examining them, to discover their points of
agreement and of difference ; particularly as,
though the order is such a very large one, the
flowers of all the plants are so much alike, that
no one can be in doubt respecting their alliance.

ORDER XIV.—RESEDACEÆ.—THE MIGNONETTE.

THE common Mignonette (*Reseda odorata*) was once included in the order Capparideæ ; but it is now made into a little order by itself, called Resedaceæ.   The flower, as is well known, is by no means remarkable for its beauty, though it is for its fragrance ; but when examined botanically, it will be found well deserving of attention, from the singularity of its construction.   It has a green calyx of six sepals, which are only remarkable for being what botanists call linear ; that is, long and narrow, and of equal width throughout—a very unusual form for sepals.   Within the calyx are the petals, six fleshy, green, heart-shaped bodies ; with a hair-like fringe round the lower part, and with the upper part cut into a tuft of segments so different in colour and texture from the lower part, that it is scarcely possible to believe that they are one.   This upper part of the petal is called the crest, and it is pure white ; the segments into which it is divided appearing to be a reat number of delicate little petals growing out of a fleshy heart-shaped disk.

It is worth gathering a flower of Mignonette, and taking off one of the petals to look at it in a microscope ; and one of moderate

size, which may be bought for 12*s.* at the
Bazaars, will be quite sufficient for the purpose.
It will then be found that the fleshy part of the
petal is as easily detached from the rest of the
flower as petals usually are, but that it is so
firmly fixed to the crest as to be inseparable
without cutting.  The disk is, however, dilated
and curiously drawn out between the stamens
which are inserted in its base, and the petals,
as though to form a barrier between them.  It
will be quite visible when the petals are removed,
as it is as firmly attached to the stalk of the
flower as the petals are to their crests.  Between this elevated part of the disk and the
calyx is a green substance which looks like a
part of the stalk, but which belongs to the disk.
There are twelve stamens, with large orange-
coloured anthers, which are at first erect, but
afterwards bend forward horizontally; and in
the middle of the stamens is the ovary, an
oblong hollow cell, with a three-lobed stigma,
forming three erect points.  Inside the ovary
from each stigma runs the kind of nerve called
the placenta, and to each nerve are attached
three rows of seeds.  The substance of the
ovary is always soft and leaf-like, even when it
becomes a ripe capsule; and though it is greatly
swelled out and bladdery, it retains the same
leaf-like and somewhat wrinkled appearance to

the last.   When the capsule is ripe, each of
the pointed lobes, which formed its upper extremity, opens down the middle, thus forming a
curious three-cornered mouth for the discharge
of the seed.   The flowers form what is called an
upright raceme,  springing from a succulent
main stem, which is, however, somewhat woody
at the base.   The plant indeed, though treated
in  England as an annual, is a shrub in the
plains of  Barbary,  of which it is a native ; and
even in this country it may be made to assume
the character of a small tree,  by keeping it
during winter in a hot-house or green-house.   I
was very much surprised to find that Mignonette has been introduced barely a century ;
and it seems difficult to imagine how those of
our ancestors who were fond of  flower-gar-
dens contrived to do without it.   I have only
to add that there are several species of Reseda,
one of which (*R. luteola*) is a British plant used
in dyeing, and is called Dyer's Weed, or Weld.

---

### ORDER XV.—DATISCEÆ.

THERE is only one genus of three species,
which are coarse hardy perennials, having the
appearance of hemp ; and only grown in England in botanic gardens.

T

ORDER XVI.—CAPPARIDEÆ.—THE CAPER TRIBE.

THIS order is divided into two sections, viz.,
the true Capers, and the Cleomes; both of
which have very long and conspicuous stamens.
The common Caper (*Capparis spinosa*) has a large
and handsome flower, with a distinct calyx and
corolla, both in four parts. The petals are
white, and so delicate in their texture as to
fade in a few hours if exposed to the sun; and
the stamens, which are very numerous, have
rich purple filaments. In the centre is the
pistil, with a very long stalk, and the ovary
at the point, instead of at the base, with no
style, and a very small stigma. In conse-
quence of this curious construction, the seed-
pod, which is fleshy, and hangs downwards,
appears to be on a much longer stalk than the
flower. The shrub is spiny, and in its natural
habitat it grows among stones and rocks. It is
the unopened flower-buds that are pickled. The
genus Cleome consists principally of annuals,
with very handsome flowers, which have very
long stamens, and a pistil of the same construction as in Capparis, but the fruit is a dry capsule. The anthers of the stamens are often
enfolded in the flowers before they are fully
expanded, so that the filaments appear bent,
till at last they open fully and hang down.

There are a few other genera in the order, but they are little known in England.

---

## ORDER XVII.—FLACOURTIANEÆ.

THE plants belonging to this order are mostly Indian plants, little known in Europe.

---

## ORDER XVIII.—BIXINEÆ.—THE ARNOTTA TRIBE.

*Bixa Orellana* is a shrub, a native of South America, which requires a bark stove in England. It has pink flowers with five petals, and a green calyx of as many sepals. The stamens are numerous; but they are on rather short filaments. The leaves are very large and heart-shaped. The fruit is a berry, and the pulp in which the seeds are immersed, when dry, is the Arnotta used in colouring cheese.

---

## ORDER XIX.—CISTINEÆ.—THE CISTUS TRIBE.

THERE are only four genera in this order, viz., Cistus, Helianthemum, Hudsonia, and Lechea; and though there are almost innumerable plants comprised in it, they nearly all belong to the first two.

All the plants belonging to the genera Cistus, the Rock-rose, and Helianthemum, the Sun-rose,

have showy flowers, each having five petals,
which are crumpled in the bud like those of the
Poppy ; they also resemble the petals of the
Poppy tribe in falling almost as soon as they
have expanded, as every one must have observed
who has noticed the flowers of a Gum-Cistus.
The calyx in both Cistus and Helianthemum
generally consists of five sepals, two of which
are larger and of a paler green than the others,
and grow a little below them ; and this calyx
remains on after the petals have fallen, and,
indeed, till the seed is ripe. In the Gum-Cistus,
however, and some other species, the two outer
sepals are wanting. There are a great many
stamens, which are rather short, and form a
tuft in the centre of the flower, surrounding the
pistil, which has a round flat-headed stigma, a
rather long style, and an ovary divided into
five cells. The seeds are numerous, and each
has a separate foot-stalk, by which it is attached
to the placentas, which, in the Cistus tribe, are
in the centre of the ovary, and not proceeding
from its sides, as in the Mignonette. The
capsule, which remains covered with the calyx
till it is quite ripe, divides into five or ten concave valves, each having a placenta, to which
the seeds were attached, in its centre. The
seed of any plant belonging to the order Cista-
ceæ, is remarkable when cut open for the great

size of the embryo enclosed in it, and the curious manner in which it is curled up. The embryo is the germ of the future plant, and it is usually buried in a great mass of albumen, or floury matter intended for the nourishment of the young plant, till its roots are in a fit state to supply it with nourishment. In the seed of the Cistus, there is scarcely any albumen ; but in its stead a long narrow embryo, coiled up like a sleeping snake.

The Gum-Cistus is generally called, in the nurseries, Cistus ladaniferus ; but it differs materially from the plant so named by Linnæus, as that has a ten-celled capsule, while the capsule of the common Gum-Cistus (which botanists call *C. Cyperius*) has only five cells. The leaves also differ, the under surface of those of the one kind being woolly, and of the other smooth; the one is also a native of Spain and Portugal, and the other, as the specific name imports, of Cyprus. Both species, and also *C. Ledon*, exude from their stems and leaves, a kind of gum or resin called Ladanum or Labdanum, which is used in medicine. It is from this gum having been formerly always mixed with opium when that drug was dissolved in spirits of wine, that the name of laudanum is given to the tincture of opium.

The two genera, Cistus and Helianthemum,

differ chiefly in the capsule, which in the latter
genus is triangular and one-celled, opening into
three valves, each of which has a narrow dissepiment down its centre.  To prevent any confusion arising from the use of these terms, I may
here observe that when a capsule is divided into
several cells, having no communication with
each other, the membranes that separate them
are called dissepiments; while the nerve-like part
of it to which the seeds are attached is named
the placenta.  Sometimes the placenta is
merely a nerve running down the side of the
capsule, when the capsule is one-celled, without
any dissepiment or division ; and sometimes the
dissepiment does not spread across the capsule
so as to divide it into different cells, but only
projects a little way from the side towards the
centre, as in the one-celled capsule of the Poppy,
(see p. 260,) and in that of the Helianthemum,
when the seed-vessel opens naturally into different parts, as in the Gum Cistus, these parts
are called valves, as are also the parts of pods,
as shown in the curled-up valves of the silique,
*fig.* 115, in p. 268.

To return to the Helianthemum, the species
of this genus are generally used for rockwork,
as they are all dwarf plants, though many of
the genus Cistus are large shrubs four or five
feet high.  The English name of the Helian-

themum, Sun rose, is very appropriate, as the
flowers will only expand in sunshine, and will
even decay in the bud without opening at all,
when gloomy weather lasts for several days.

---

ORDER XX.   VIOLACEÆ—THE VIOLET TRIBE.

THE order Violaceæ, though not a large one,
contains several genera, but the most interesting
is the genus Viola, which includes among many
other species the Sweet Violet (*Viola odorata*),
and the Heartsease (*V. tricolor*).   The flowers
of both species have many claims to admiration,
but they do not add the charm of regularity in
construction to their other attractions, as, in
fact, few flowers are less symmetrical.   The
flowers of both are nearly alike in their details;
but to avoid confusion, I will describe them
separately.   The calyx of the Heartsease consists of five pointed distinct sepals, two of them
rather smaller than the others.   These sepals
are not attached, as in most other plants, at
their base ; but so as to leave nearly a quarter
of their length standing up, far beyond the
place where they are fixed to the receptacle,
so as to form a sort of border or cup round
the stem, and between it and the flower.
The sepals are green, but they are edged
with a delicate whitish membrane at the margin, scarcely to be seen without a microscope.

There are five petals which are also irregular
in their construction, two of them being much
larger than the others, and generally of a different colour ; and one even of the other three
being quite different in form to its companions.
The two large petals at the back of the flower,
which in the common Heartsease are generally
dark purple, are laid over each other, and behind the two below them.   These two side-
petals, which form the centre of the flower, are
both furred at the base ; and the lower petal,
which is placed between them, has its claw
drawn out behind into a spur, which passes
between two of the sepals ; and which, when
the flower is looked at from behind, appears to
be part of the calyx.   The furred part of the
two side-petals forms a triangular, roof-like
opening, peeping out of which, is seen a small
pale-green ball-like substance, which a fanciful
imagination might liken to a head looking
through a dormer window ; and this is all that is
to be seen in place of the usual apparatus of
stamens and pistils.   As all seed-producing
flowers must have stamens and pistils, and as it is
well known that Heartseases and Violets do produce seed in abundance, it is clear that these
important organs are not wanting ; but where
are they ?   It is easy to guess, after being so
far initiated in the mysteries of botany, that the
little globular body is a part of the pistil ; but

where are the stamens? It is necessary to pull
the flower to pieces to discover them. Commencing this work of destruction, which I
always feel remorse at perpetrating, for I love
flowers too well not to feel pain at destroying
them; commencing this work, I repeat, the
petals and the sepals must be carefully removed
from the stem; a task of some little difficulty,
as both sepals and petals are firmly attached to
the receptacle, and the lower petal must have
its spur opened with a pin to avoid hurting the
delicate organs it contains. When the outer
coverings of calyx and corolla are thus both
removed, the seed-producing organs will be discovered, and it will be found that they consist
of five very curiously-formed stamens, with as
singular a pistil, in their centre. The stamens
have no apparent filaments, and the anthers,
which seem to be inserted in the receptacle,
look like seeds, each tipped with a bit of brown
skin, and having what appears to be a white rib
in front. This rib is the anther; and the
broader part is the dilated filament, which is
drawn out beyond it, on both sides, and above,
so as to form the brown tip above the anther
already mentioned. Two of the anthers have
each, in addition to these peculiarities, a long
tail, which the spur of the lower petal concealed,
when the flower was in a perfect state. The

pistil consists of a large ovary, full of ovules,
with a narrow style, which is drawn out into
the hollow globular termination which is seen
through the triangular opening in the flower.
The globe has an opening in front, under which
is a kind of lip, which looks like a shutter let
down to show the opening ; and though, from
its thick fleshy nature, it looks like a stigma, it
is only the outer covering of that organ, for the
stigma lies within the opening.   In this manner
the stigma and anthers are completely concealed;
and thus it will be seen, that nothing can be
more complex and intricate than the construction of the flowers of the Heartsease.

Who could suppose that all these elaborate
details would be necessary to illustrate so simple a flower as that of the Violet?   And yet the
construction of the flowers of the Violet and
those of the Heartsease are essentially the same.
The sepals of the Violet are extended at the
base, like those of the Heartsease, and the
corolla consists of the same number of petals,
which are equally irregular in their form,
though not in their colour ; the lower petal is
drawn out, in the same manner, into a spur,
which is much longer than that of the Heartsease, though the rest of the flower is smaller.
The stamens are formed with the same regular
irregularity, only the tails of the two irregular

ones are larger and stronger, in the same proportion as the spur is larger which is intended
to conceal them.  The pistil is of the same
shape, with the same curiously constructed and
perforated style, which is bent in its narrow
part and swelled out into a hollow globe at the
tip ;  and in both species, the ovary is one-celled
with three parietal placentas, that is, with three
nerve-like projections from the sides of the
capsule, having four rows of seeds attached to
each.   The capsule looks like a smooth shining
berry, and it remains partially shrouded by the
calyx, till the seeds are ripe ; when it bursts
open with an elastic spring, and divides into
three valves, each of which has the placenta
bearing the seeds in its centre.

In all these points the Heartsease and the
Violet are alike ; but they differ materially in
the leaves, which in the Violet are broad and
heart-shaped, without stipules ; but in the
Heartsease are small and ovate, with such very
large and deeply-cut stipules, that they are by
most persons mistaken for the leaves.  I may
here be asked what are stipules, and in what
do they differ from leaves ?  In answer to the
first question, I can only inform my readers that
stipules are generally little leaf-like bodies,
which seem to act as attendants upon leaves, as
bracts seem to wait upon flowers ; but in what

they differ from leaves, except in size and shape,
I have not been able to learn.  Even Dr. Lindley
in the last edition of his Introduction to Botany,
says, " What stipules really are is not well made
out."

The Heartsease and the Violet differ also in
their habit of growth.  The Violet is a creeping
plant with no stalks but those supporting the
flowers, while the Heartsease stands erect, with
a thick square stem, so strong, that, notwithstanding its succulent nature, it be may trained
like a little tree.

--------

ORDER XXI. DROSERACEÆ—THE SUN-DEW TRIBE.

THERE are three genera in this tribe that are
well known: *Drosera*, the Sun-dew; *Dionæa muscipula*, Venus's Fly-trap; and *Parnassia palustris*,
the Grass of Parnassus; all bog plants.   The
species of the genus Drosera are remarkable
for the curious manner in which the leaves and
peduncles are coiled up when they first appear,
and in which they slowly unroll themselves as
they grow.  They are also beautifully edged
with a sort of fringe of glandular red hairs, and
a fluid exudes from these glands which makes
them always appear as though covered with
dew.   The common Sun-dew (*D. rotundifolia*)
is a British plant, with short roundish leaves;

but other species are natives of New Holland
and North America ; and several of them have
long slender leaves like threads.   Venus's Fly-
trap (*Dionæa muscipula*) is a native of Carolina, in North America ; the leaves are curiously
formed of two lobes, which close and open as
if hinged, and they are furnished with glandular
hairs, which are so extremely irritable as to
make the leaves close at the slightest touch, and
thus to imprison any unfortunate insect that
may be within the lobes.   The petiole is so
much dilated as to look like a leaf, but the real
leaf consists of only the two roundish lobes
edged with teeth that form the Fly-trap.   The
flowers are white, and they are produced in
corymbs.   The corolla has five petals, which do
not fall off when they wither, but roll up so as
to look like the cocoon of an insect.

ORDER XXII.   POLYGALEÆ—THE MILKWORT
TRIBE.

The genus Polygala is well known from the
very handsome greenhouse plants which it contains.   The flowers at first sight appear to
resemble those of the Sweet Pea, having two
wings like a standard, and a sort of keel ; their
construction is, however, very curious, and so
complicated, as to be very difficult either to

describe or to understand. The calyx is said
by modern botanists to consist of five sepals,
three of which are green and two lilac, these
last being the part that resembles the standard
of the Sweet Pea. The corolla is also said to
consist of five petals, two of which stand erect,
and the other three grow together to form the
keel. The latter have their upper part cut into
a kind of crest, like that of the Mignonette.
Below the crest, the united petals form a kind
of hood, under which are arranged the eight
stamens, four on each side. The stamens
themselves are as remarkable as the other parts
of the flower; the filaments grow together into
a thin kind of leaf, and each anther has but one
cell, and opens by a pore at the apex. The
pistil is also very curiously formed, as the style
and stigma have the appearance of a gaping
monopetalous corolla. The fruit is a flat two-
celled capsule, which, when ripe, opens by two
lips, separating from each other, and showing a
seed within each cell. Even the seeds are not
like other seeds, for each has a large white protuberance at one end, called a corancula.

---

### ORDER XXIII. TREMANDREÆ.

SLENDER New Holland shrubs, with the habit
of Heaths, rarely met with in British gardens.

ORDER XXIV.　PITTOSPOREÆ.—THE PITTOSPORUM
TRIBE.

THE principal genera included in this order
are Pittósporum, Billardiera and Sollya, all
resinous shrubs, with alternate leaves without
stipules, and the sepals and petals, each five in
number, and laid over each other like scales in
the bud. The seeds are numerous, and immersed in fibrous pulp. The commonest species
of Pittosporum is *P. Tobira,* a native of China,
easily known by its thick leathery leaves, the
mid-ribs of which are strongly marked, and
whitish. The flowers are erect, and produced
in cymes or heads ; and the petals are united
into a tube with a spreading limb. The capsule
is one-celled, and two or three valved, with an
imperfect dissepiment in the centre of each
valve ; and the seeds are numerous, and buried
in a resinous fibrous pulp. The Billardieras are
generally climbing shrubs, with pale greenish
bell-shaped, and almost erect flowers, which are
produced singly or in pairs, and which have the
tips of their petals turned back. The fruit is
a fleshy berry, with a shining skin of a deep
blue, and it is called the Apple Berry in
Australia, of which country the species are
natives. This fruit is said to be eaten in
Australia, but it seems difficult to imagine how

this can be the case; as though the outer part
of the berry is of a soft spongy nature, it is dry
and insipid; and there is no internal pulp, for
the seeds lie loose in the cells.   In *Sollya
heterophylla* the flowers are drooping, on long
and very slender pedicels, and they are produced
in cymes.   The corolla is campanulate, with
the tips of the petals not recurved, and the
anthers are much shorter than in Billardiera.
The fruit is a soft fleshy berry, divided into two
cells, each containing two rows of seeds immersed in pulp, and when cut open, it smells
strongly of turpentine.    The plant generally
called *Sollya linearis* has a dry and leathery
pericardium; and for this reason and on account
of the spreading of its anthers, it was placed by
Mr. Cunningham in a new genus, which he
called *Cheiranthera*.

---

## ORDER XXV.   FRANKENIACEÆ.—THE FRANKENIA
## TRIBE.

The genus Frankenia consists principally of
the British weeds called Sea Heath; and the
other genera included in the order are seldom
seen in British gardens, from the seeds which
have been imported seldom arriving in a state
fit for vegetation.

ORDER XXVI.—CARYOPHYLLACEÆ—THE CARNA-
TION TRIBE.

THE plants belonging to this order have so
strong a family likeness to each other as to be
easily recognised; and they are all distinguished
botanically by the swollen joints of their stems,
and their opposite undivided leaves, which are
generally connate, that is united, and sheathing
the stem.   The order is divided into two sections, viz. : Sileneæ, in which the sepals are
united into a tube, and which section includes
the genera   Silene,   Dianthus,   Saponaria,
Lychnis, and Agrostemma ; and Alsineæ, in
which the sepals are either quite distinct, or
only slightly cohering at the base, and which
includes Stellaria, Arenaria, Cerastium, Sper-
gula, and several other British weeds.   The
Chickweed was   called   by   Linnæus  *Alsine
media*, but the genus Alsine is now united to
Stellaria.

The Wild or Clove Carnation (*Dianthus
Caryophyllus*), which may be considered the
type of the order, has an erect stem, swollen at
the joints, with connate leaves, (see *a* in *fig.* 118).
The flower, when single, consists of five petals,
each with a very long narrow claw (*b*), and a
rather broad limb or blade (*c*) serrated at the
edge.    The calyx (*f*) is tubular, with five

vandyked teeth, which are in fact the tips of
five sepals, into which the tube of the calyx may

FIG. 118.—THE CARNATION.

be easily divided with a pin.   The tubular form

of the calyx is admirably contrived to support
the long claws of the petals, and to keep them
in their proper places; particularly when the
flowers are double, as the weight of the petals
in that case frequently bursts the tube of the
calyx. Every one fond of pinks and carnations
must have observed the miserable appearance
of the flower when thus deprived of its natural
support; and to prevent the premature destruction of prize-flowers by this misfortune, pro-
fessed florists sometimes slip a curiously-cut
piece of card-board over the bud, which remains
on after the expansion of the flower, and pre-
vents the petals from falling out of place. Some
florists tie the calyx round with thread, instead
of using a pasteboard ring, which answers the
same purpose. At the base of the calyx are
two, four, or six leafy appendages (g), resembling bracts, which are called the calycine
scales. These imbricated scales are, however,
only found in the genus Dianthus. There are
ten stamens (d) unequal in height, but none of
them longer than the ovary round which they
are placed. The ovary and the stamens are
concealed in the cup of the flower, but the
former is furnished with two styles, terminating
in two long stigmas (e), which project beyond
the flower, and which, when magnified, appear
delicately fringed.

The genus Dianthus includes the Carnation, the Pink (*Dianthus plumarius*), the Chinese Pink (*D. sinensis*), the Sweet William (*D. barbatus*), and many ornamental flowers. Of these the Sweet William has the claws of its petals bearded; the flowers are produced in bundles or fascicles; and the calycine scales are so numerous and awl-shaped, that they give a bristly appearance to the flowers. The different species of Soap-wort (*Saponaria*) differ from Dianthus, in having no calycine scales; and this is also the case with the berry-bearing Campion (*Cucubalus baccifer*), the fruit of which is a fleshy capsule or berry, which finally becomes black, and has a singular appearance in the centre of the cup-like calyx, which remains on till the fruit is ripe. The flower of this plant is white, and the petals have a two-cleft limb. All the numerous species of Catchfly (*Silene*) are also without calycine scales, and the petals are generally deeply two-cleft; but they are distinguished by having a crown of petal-like scales in the throat of the corolla. There are also three styles instead of two; and the capsules are three-celled at the base, ending in six teeth at the top. The species have frequently a glutinous frothy moisture on the stem, in which flies sometimes become entangled, and hence the English name of the genus. One species, the

Bladder Campion (*S. inflata*), has been used as
food, and its young shoots, when boiled and
sent to table like Asparagus, are said to have
the flavour of green peas. The different species
of Lychnis and Agrostemma resemble Silene
closely in every respect, except in the styles,
which are five, instead of three; these two
genera, Viscaria, and Githago, differ very
slightly from each other; and several of the
species are known by different names : thus
Ragged Robin (*Lychnis flos cuculi*) is made by
some botanists an Agrostemma; the Corn-cockle
is sometimes called *Githago segetum*, and sometimes *Agrostemma Githago;* the common Rose
Campion is called sometimes Lychnis, and
sometimes Agrostemma; and the Rock Lychnis,
or Red German Catchfly, sometimes *Lychnis
Viscaria*, and sometimes *Viscaria vulgaris*.

---

ORDER XXVII.—LINACEÆ.—THE FLAX TRIBE.

The order Linaceæ is a very small one ; and,
indeed, it consists principally of the genus
Linum. The Flax was formerly included in the
Caryophyllaceæ, which it resembles in having
five petals, five sepals, and five stamens ; but it
also resembles the Mallow in its capsules, and
in its stamens growing together at the base;

and the Cistus in its persistent calyx, and the
disposition of its sepals. These links, which
connect one order with another, and make them
appear alike but not the same, form, I think,
one of the most interesting parts of the Natural
System. We are led on from one gradation to
another, by scarcely perceptible shades of difference through the vegetable kingdom; and,
indeed, through the whole system of creation:
the beautiful harmony, and unity of design,
visible throughout, bearing the strong impress
of the Divinity whose power has made the
whole.

The common Flax (*Linum usitatissimum*),
though in its appearance only an insignificant
weed, is a plant of great benefit to man. The
fibres of the stem are used to make linen, and
the seeds (linseed) are crushed for oil. The
flowers are blue, and have five regularly-shaped
petals, which are twisted in the bud ; and a
distinct calyx of five pointed sepals, two of which
grow from a little below the others, as in the
Gum Cistus; and, as in that plant, the calyx
remains on till the seeds are ripe. There are
five stamens, the filaments of which grow together slightly at the base, and there are five little
points like filaments without anthers, rising
between the stamens. The petals are connected
with the ring formed by the united filaments,

and sometimes the petals themselves grow
slightly together at the base. The capsule
consists of five two-celled carpels, grown together ; each cell containing one seed, and each
carpel terminating in a rather slender style,
tipped with a ball like stigma. When ripe, the
capsule opens naturally, by dividing into ten
valves, to discharge the seeds ; which are flat
and shining, with a large embryo. These seeds
are called Linseed in the shops, from Linum,
the botanical name of the plant; and, as is well
known, they are not only used for various purposes, but oil is expressed from them. The
stem of the common Flax, though it is only an
annual, consists of woody fibre, like that of a
tree in its young state ; and it is this fibrous
part that makes the yarn for thread, after it
has been separated from the fleshy part, by
steeping the stems for a long time in water.
The perennial Flax (*Linum perenne*), which, as
its name imports, lasts several years, differs in
little else from the common kind, except that
its sepals are obtuse, and its leaves are much
smaller and narrower. Both these are natives
of Britain. There are many other species,
some of which have yellow flowers.

ORDER XXVIII.—MALVACEÆ.—THE MALLOW TRIBE.

ALL the plants belonging to Malvaceæ bear
so much resemblance to each other, that this
order may be considered a very natural one ;
and it is one very remarkable for the botanical
construction of its flowers.   In some respects it
resembles Linaceæ, quite enough indeed to show
clearly the chain by which they are so beautifully linked together ; but in others, it differs
so decidedly as to show how completely they
are distinct.   *Fig.* 119, which represents the

flower and seed-vessel
of the Althæa frutex
(*Hibiscus   syriacus*),
will serve to show the
chief peculiarities of
this order.   The calyx
*a* consists of five sepals,
below which is an in-
volucrum of six or
seven leaflets, which
have the appearance

FIG. 119.—THE FLOWER, STAMENS,
AND PISTILS OF THE ALTHÆA FRUTEX.

of a second calyx.   The corolla is cup-shaped,
and consists of five petals, which are close together at the base, and this is peculiar to the
genus Hibiscus.    The capsule is round and
somewhat convex, being nearly in the shape of
what is called a batch-cake,   as shown at *c* ;

it consists of five carpels grown together, each
containing many seeds ; and when ripe, it bursts
naturally into five valves, each of which has a
dissepiment down the centre. The filaments
grow together very curiously, inclosing the
styles, and forming a column in the centre of
the flower, which is the distinguishing mark of
the Malvaceæ. Some of the stamens are shorter
than others, and as part of each filament is detached, the anthers form the fringe-like border to
the column, shown at *d.* The anthers are kidney-shaped and one-celled, and this is another
of the characteristics of the order ; but the
styles are terminated by five ball-shaped stigmas,
like those of the Linum. There are many
kinds of Hibiscus ; but perhaps the best known
are : *H. rosa sinensis,* the species which is so
often represented in Chinese drawings, and the
petals of which are so astringent, that they are
said to be used in China by the men to black
their shoes, and by the women to dye their
hair ; and the Bladder Ketmia (*H. Trionum*),
which takes its English name from its inflated
capsule. All the plants belonging to the order
Malvaceæ have a central column, round which
are placed numerous carpels, which grow together and form a many-celled capsule ; and they
all have kidney-shaped, one-celled anthers.
They have also always an involucrum below the

calyx, but this involucrum differs in the different
genera. In the genus Malva, the involucrum
consists of three leaflets, which in the common
Mallow (*Malva sylvestris*) are oblong. The
petals are wedge-shaped, and they are what
botanists call auricled; that is, they are set so
far apart at the base that light can be seen
through them. The stamens are all of nearly
the same height, and they form a kind of bunch
round the styles, which are pointed. The capsule consists of a circle of woolly-looking carpels
growing close together, but so as to be easily
detached with a pin, and each fitting into a
little groove in the receptacle, in which they are
placed. As the seeds ripen, the involucrum falls
off, but the large loose-looking calyx remains
on. There is only one seed in each carpel; but
as there are generally eleven carpels in each
capsule, each seed-vessel contains this number
of seeds. The leaves are lobed and toothed;
and the whole plant is covered with long hairs,
which are disposed in little star-like tufts.

The genus Malope closely resembles the Mallow; except that the petals are not wedge-
shaped, and that it has a still larger calyx, the
long sepals of which shroud the capsule as the
involucre of the filbert does the nut. The involucrum is composed of three broad, heart-shaped
leaflets, which remain on till the seed is ripe.

The petals are also not so even along the margin ; and the carpels are so disposed as to form a cone-shaped capsule, instead of a flat one.

The genus Lavatera has the leaflets of the involucre joined to the middle, so as to form a kind of three-cornered saucer below the capsule ; and the capsule itself is completely covered with a part of the receptacle, which is dilated, and curved down over it.    Lastly, the genus Althæa, the Marsh Mallow, has the involucrum cleft into six or nine divisions, and the carpels united into a globular capsule.    The Hollyhock (*A. rosea*) belongs to this genus.    Many other genera might be mentioned, but these will suffice to give my readers a general idea of the order, and of the points of difference which distinguish one genus from another.    Among the exotic plants belonging to the order is the cotton tree (*Gossypium herbaceum*), the cotton being the woolly matter which envelops the seeds in the capsule.    All the Malvaceæ abound in mucilage, and they all have woody fibre in their stems.

---

### ORDER XXIX.—BOMBACEÆ.—THE SILK COTTON TREE TRIBE.

THIS order is closely allied to Malvaceæ, and it differs principally in the tube formed by the stamens being divided into five bundles near the top.    It includes the Baobab, or Monkey-bread

(*Adansonia digitata*), said to be the largest tree
in the world; the Screw tree (*Helicteres Isora*),
so named from its curiously-twisted fruit; *Caro-
linia princeps;* the Silk Cotton tree (*Bombax
Ceiba*); and the Hand-plant (*Cheirostemon platanoides*),—this is the Hand-plant so named from
the lobes of its leaves resembling fingers,—all
stove plants in Britain.

---

ORDER XXX.—BYTTNERIACEÆ.—THE BYTTNERIA
TRIBE.

This order is divided into five sections, which
some botanists make distinct orders. It is very
nearly allied to Malvaceæ, but the anthers are
two-celled. The principal genus in the first
section (*Sterculieæ*) is Sterculia, which has
several carpels distinct and arranged like a star:
the species are trees with large handsome leaves
which are articulated at the base, and axillary
panicles or racemes of flowers. The second section
(*Byttnerieæ*) contains among other plants *Theo-
broma Cacao*, from the fruit and seeds of which
Cocoa and Chocolate are prepared. The third
section (*Lasiopetaleæ*) is well known in England,
by the pretty Australian shrubs included in the
genera Thomasia and Lasiopetalum, the leaves
of which have their under surface downy, and
generally brown. The fourth section (*Herman-
nieæ*), and the sixth (*Wallichieæ*), contain no

plant common in English gardens ; and the
fifth (*Dombeyaceæ*) is best known by *Astrapæa
Wallichii*. The qualities of all the plants in this
order are mucilaginous.

---

### ORDER XXXI.—TILIACEÆ.—THE LINDEN TRIBE.

THE only genus belonging to the natural
order Tiliaceæ which is easily to be procured
in Britain is that of Tilia, the Lime trees. The
common Lime (*Tilia europæa*) is generally a
tall, well-formed tree, with rather broad leaves,
which are heart-shaped at the base, tapering at
the point, and serrated at the margin : they are
also smooth on the outer surface, thin, and of a
light and delicate texture ; below there is a
little tuft of hair at the angle of the veins.   The
flowers are produced in cymes or compound
umbels (see *fig.*
120) ; and their
main pedicel ap-
pears to spring
from one long en-
tire    bract   (*a*).
The   calyx is in
five sepals, and it
falls off before the
corolla, which is
composed of five

FIG. 120.—THE FLOWERS AND SEED-
VESSEL OF THE COMMON LIME TREE.

pale yellow petals, which are very sweet-scented.

The stamens are numerous, and the filaments
separate, bearing two-celled anthers, which burst
by long slits.  The ovary has only one style, the
tip of which is cleft into five small stigmas;  and
it is divided into five cells, each containing one
or two ovules.  The fruit or capsule (*b*) is round,
and has a leathery skin, covered with a soft
down;  and when ripe, the cells often become
united so as to form one, with only one or two
perfect seeds in the whole capsule, the other
ovules proving abortive.  The whole plant
abounds in mucilage, and the sap when boiled
affords sugar.  The inner bark is so tough and
fibrous, that it is used for making what are
called bast mats: it being first rendered flexible by steeping it for a long time in water.  The
wood is of very fine texture, but soft and white,
and it is thus admirably adapted for carving.
The American Limes have a small scale at the
base of each petal of the flower;  but the other
differences between the species are very slight.

ORDER XXXII.—ELÆOCARPÆ.—THE ELÆOCARPUS
TRIBE.

East India shrubs and trees, little known in
Britain.  " The hard and wrinkled seeds of
Elæocarpus are made into necklaces in the
East Indies, and, set in gold, are sold in our
shops."—(*Hook.*)

ORDER XXXII*.—DIPTEROCARPÆ.—THE CAMPHOR
TREE TRIBE.

THERE are two kinds of Camphor, one produced by boiling the branches of a kind of laurel,
and the other (the Camphor of Sumatra) is
found in large pieces in the hollow parts of the
branches of *Dryobalanops Camphora*, one of the
species included in this order.  None of these
trees have been introduced into Britain.

ORDER XXXIII.—CHELONACEÆ, OR HUGONIACEÆ.

SMALL trees and shrubs, natives of the East
Indies and Madagascar; only the genus Hugo-
nia is known in Britain.

ORDER XXXIV.—TERNSTRŒMIACEÆ.—THE
TERNSTRŒMIA TRIBE.

THE principal plants in this order common in
England are Gordonia, Stuartia, and Malachodendron.  *Gordonia Lasianthus*, the Loblolly
Bay, is a small evergreen tree, with white
flowers, about the size of a rose.  It is a native
of America; and Stuartia and Malachodendron
are beautiful low trees or shrubs, with large
white flowers from the same country.  The
flowers have five large petals; the stamens are

numerous, with the filaments growing together
at the base, and attached to the petals; and
there are five carpels more or less connected.
Gordonia has its five sepals leathery, and
covered with a silky down; its stamens almost
in five distinct bundles, a five-celled capsule,
and its seeds each furnished with a wing.
Stuartia has a permanent calyx, five-cleft, but
not parted into distinct sepals, with two bracts
at the base, and a woody five-celled capsule,
with seeds without wings; and Malachodendron
(which was formerly called *Stuartia pentagynia*)
has a calyx similar to that of Stuartia, but the
edges of the petals are curiously crenulated, and
there are five distinct carpels, each containing
only one seed. Some botanists include the
Camellia and the Tea in the order Ternstrœmiaceæ.

---

### ORDER XXXV.—CAMELLIACEÆ.—THE CAMELLIA TRIBE.

THERE are two genera in this order, the
Camellia and the Tea. The flower-bud of the
Camellia is inclosed in a calyx of five, seven, or
nine concave sepals, on the outside of which are
several bracts, which remain on till the flower
has expanded, but which are distinguished from
the sepals by their dark brown colour. The
sepals and the bracts are laid over one another

like scales, and thus the flower lies encased in a complete coat of mail. The single flower is cup-shaped, with five, seven, or nine petals, which are sometimes joined together at the base. The stamens have long slender filaments, which either grow together at the base, or are separated into several bundles. The anthers are elliptical and versatile; that is, they are poised so lightly on the filament as to quiver with the slightest breeze. The ovary is of a conical shape, and it has three or five slender styles, ending in as many pointed stigmas, and growing together at the base. The capsule is three or five-celled; and when ripe it bursts into three or five valves, in the middle of each of which is a dissepiment, which, before the capsule opened, was attached to an axis or column in the centre. The seeds are large and few, and they are fixed to the central placenta. There is no albumen, but the embryo has two large, thick, oily cotyledons, which look as if they were jointed at the base. The leaves are leathery, dark-green and shining, and they are ovate in form, ending in a long point, and sharply serrated. The flowers spring from the axils of the leaves, and grow close to the stem without any footstalk; and the leaf-bud for the ensuing shoot grows beside the flower-bud.

I have above described the *Camellia japonica*,

x

from which nearly all the Camellias in British
gardens have sprung; but there are some other
species. The finest of these is *C. reticulata*,
which has very large, loose, widely-spreading
flowers, of a remarkably rich crimson. The
leaves are oblong, flat, and reticulately veined,
being of a much finer texture than those
of *C. japonica*. The ovary is two or four-
celled, and it is covered with fine silky hairs.
*C. maliflora* is a very beautiful species with small
semi-double flowers, coloured like an apple blos-
som. This Camellia is by some botanists
thought to be a variety of *C. Sasanqua*, an
elegant species with white fragrant flowers;
but the ovary of the first is smooth, and that
of the second covered with hairs, which most
botanists consider a specific difference.

The Tea tree (*Thea viridis*) is very nearly
allied to the Camellia; but there are many points
of difference. The flower of the Tea tree has a
footstalk; the calyx has only five sepals; the filaments of the stamens do not grow together; the
capsules are three-seeded; and the dissepiments
are formed by the edges of the valves being bent
inwards, instead of being attached to a central
axis. The leaves are also much longer than
they are broad, and they are of a thinner texture and pale green; and the outside of the
capsule, which is furrowed in the Camellia, is

quite smooth in the Tea tree. It is said that both the green and the black Tea are made from the leaves of *Thea viridis;* but there is another species called *Thea Bohea,* which has smaller leaves, and is a more tender, and less vigorous-growing plant. The young leaves of *Camellia Sasanqua,* and some of the other Camellias, are also dried, and mixed with the tea. All these plants are natives of Japan and China, and require a slight protection in England during winter.

ORDER XXXVI.—OLACINEÆ.—THE OLAX TRIBE.

EXOTIC trees from the East and West Indies, little known in Britain. *Heistria coccinea,* a native of Martinique, is said to be the Partridge wood of the cabinet-makers.

ORDER XXXVII.—AURANTIACEÆ.—THE ORANGE TRIBE.

THE natural order Aurantiaceæ contains fourteen genera; but the only one I think my readers will feel an interest in is the genus Citrus. This genus comprises, among several other species, *C. medica,* the Citron; *C. Limetta,* the sweet Lime; *C. Limonum,* the Lemon; *C. Paradisi,* the Forbidden fruit; *C. decumana,* the Shaddock; *C. Aurantium,* the Sweet Orange;

and *C. vulgaris*, the Bitter or Seville Orange.
to these may be added *C. nobilis* the Mandarin
Orange, the fruit of which is reddish, and which
parts naturally from its rind, which is sweet,
and may be eaten. All the species agree in
having a tube-like calyx, scalloped into five
short teeth, and a flower of generally five
fleshy petals, (see *a* in *fig.* 121), though the

Fig. 121.—Flower and seed of the Orange.

number occasionally varies from four to nine.
These petals are elliptic in shape, concave, and
always widely opened. In the centre of the
flower are the stamens, varying from twenty
(which is the ordinary number) to sixty; the
anthers are two-lobed, and oblong, and the
filaments are somewhat thickened at the base,
and united there into several small bundles (*b*),
but free above. The pistil has a somewhat
globular ovary, with a cylindrical style, termi-

nating in a stigma, which is slightly raised in
the centre. The disk in which the stamens
are inserted, forms a ring round the ovary.
The fruit (*fig.* 122), which is considered by

FIG. 122.—LEAF AND FRUIT OF THE ORANGE.

botanists to be a kind of berry, is in fact a seed-
vessel with numerous cells, divided by dissepiments and a central placenta (*a*); the cells being
the quarters of the Orange, the dissepiments
the divisions between them, and the placenta
the central pith. When the flower first expands, the ovary, if cut open and examined,
will be found to be divided into several cells,
each containing two rows of ovules. As in the
preceding genera, however, many of these ovules
become abortive; and as the cells fill gradually
with cellular pulp, the seeds become detached
from the placenta, and buried in it. The seeds
themselves are very interesting; they are
covered with a thick wrinkled skin, and they
show distinctly the hilum (*c* in *fig.* 121), the
chalaza (*d*) and the raphe or connecting cord
between them, parts which are seldom to be
distinguished in seeds with the naked eye.

The leaves, calyx, and petals of the Orange,
if held up to the light, appear covered with
little dots. These dots are cells, covered with a
transparent membrane, and filled with a kind
of oil, which is exceedingly fragrant. The rind
of the fruit is covered with similar cells, filled
with a pungent oily liquid. The leaves are
smooth and shining; and they are articulated;
that is, they can be separated from the petiole
or footstalk without lacerating them. In most

of the species, the petioles are winged; that is, they are dilated into little leaves on each side (see *d* in *fig.* 122). The different species vary chiefly in the number of stamens, the thickness of the rind, the shape of the fruit, and in the wings of the petioles. In the Citron these wings are wanting entirely, and instead of them there are spines in the axils of the leaves; there are generally forty stamens, and the rind of the fruit is very thick. In the sweet Lime, the petioles are are slightly winged, and there are about thirty stamens; the fruit is small and round, with a slight protuberance at one end like that of the Lemon, and the pulp is sweet. In the Lemon the petioles are somewhat winged, the flowers have about thirty stamens; the fruit is oblong, with an acid pulp, and a thin rind. The Sweet Orange has winged petioles, about twenty stamens, and a fruit with a thin rind and sweet pulp; and the Seville Orange differs principally in having a thicker rind and bitter pulp. The China, St. Michael, and Malta Oranges, with many others, are all varieties of the Sweet Orange (*Citrus Aurantium*); and there are many other species, which I have not thought it necessary to describe.—All the species above-mentioned are natives of Asia, and most of them of China, but they have been so long cultivated in Europe and America, as to have become almost naturalised.

ORDER XXXVIII.—HYPERICINEÆ.—THE
HYPERICUM TRIBE.

THE genus Hypericum, or St. John's Wort,
agrees with the orange in having its leaves full
of transparent cells; but these cells are filled
with a yellow, resinous juice, resembling gamboge in its medicinal properties, and having a
very disagreeable smell.  There are five petals
in the corolla; and the calyx consists of five
sepals, which are unequal in size and shape, and
joined together for only a short distance.  Like
the orange the filaments grow together at the
base, in separate clusters or bundles; but in the
Hypericum these clusters are so perfectly distinct, that the stamens may be readily separated
into three or five bundles (according to the
species), by slightly pulling them.  The capsule
is dry, and of a membrane-like texture, and it
consists of three or five carpels, containing many
seeds, and each having a separate style, and a
pointed stigma.  The flowers are very showy,
from their large golden yellow petals and numerous stamens.  The genus Androsæmum, the
Tutsan, or Park-leaves, has been separated from
Hypericum on account of its fruit being one-
celled and one-seeded, with a fleshy covering,
which yields a red juice when pressed.  *H.
calycinum*, with large yellow flowers and five

tufts of stamens, is the handsomest species ; but
*H. perforatum* is the true St. John's Wort,
which the country people used formerly to
gather on midsummer eve, as a preservative
against witchcraft.

---

## ORDER XXXIX.—GUTTIFERÆ.—THE MANGOSTEEN TRIBE.

THE only genus in this order that contains
plants interesting to the English reader is
Garcinia ; and the most remarkable species are
*G. Mangostana*, the Mangosteen, said to be the
most delicious fruit in the world, and *G. Cam-
bogia*, the tree producing the gamboge, which
is a kind of gum that oozes out from the stem.
Both are natives of the East Indies.

---

## ORDER XL.—MARCGRAAVIACEÆ.

EXOTIC shrubs, mostly natives of the West
Indies, with spiked, or umbellate flowers, and
alternate leaves. Very seldom seen in Britain.

---

## ORDER XLI.—HIPPOCRATACEÆ.

EXOTIC arborescent, or climbing shrubs, generally with inconspicuous flowers. Natives of
the East and West Indies.

## ORDER XLII.—ERYTHROXYLEÆ.—THE RED WOOD TRIBE.

EXOTIC shrubs, and low trees, remarkable from the redness of their wood, but with small greenish flowers. The leaves of *Erythroxylon Coca* possess an intoxicating quality, and are chewed by the Peruvians, in the same manner as the Turks take opium.

---

## ORDER XLIII.—MALPIGHIACEÆ.—THE BARBADOES CHERRY TRIBE.

SEVERAL species of Malpighia, the Barbadoes Cherry, are found occasionally in our stoves. The corolla of these plants, when closed, bears considerable resemblance to that of a Kalmia ; but the flower when expanded is more like that of a Clarkia, from the long claws of the five petals, and the distance they are placed apart. Several of the species have their leaves and stems beset with stinging bristles, which adhere to the hands when touched. The fruit, which is eatable, but insipid, is a berry-like drupe, containing three one-seeded nuts. The species are natives of the West Indies, and they require a stove in England. The flowers are generally rose-coloured or purplish; but they are sometimes yellow. The common Barbadoes Cherry is called *M. glabra*, and its leaves are without stings.   In Hiptage, another genus of this order, four of the petals of the flowers are white, and one yellow; and in Banisteria, the species are generally climbing shrubs, always with yellow flowers.   Some of the species of Banisteria are occasionally found in stoves in this country, where their beautiful feathery yellow flowers have very much the appearance of those of the Canary bird flower (*Tropæolum peregrinum*).

---

ORDER XLIV.—ACERINEÆ.—THE MAPLE TRIBE.

THE common Maple (*Acer campestre*) and the Sycamore (*A. Pseudo-Platanus*) are the only plants belonging to this order, that are natives of Britain; though so many kinds of ornamental Acers are now found in our parks and pleasure-grounds.   Few trees are indeed more deserving of culture than the American Maples, both for their beauty in early spring, and for the rich shades of yellow and brown which their leaves assume in autumn.   The Maple tribe is a very small one; it consists indeed of only the genera Acer and Negundo, and an obscure Nepal genus, of which there are no plants in Britain.   Of all the Acers, one of the handsomest is  the  Sycamore  tree  (*A.*

*Pseudo-Platanus*) ; the flower of this species (see
*e* in *fig.* 123) is of a yellowish green ; and as in
early spring, when it appears, we are delighted

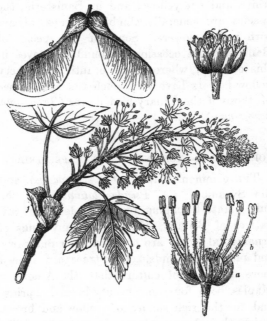

123.—FLOWER AND SAMARA OF THE SYCAMORE. (*Acer Pseudo-
Platanus.*)

at the sight of any thing in the way of flowers,
it really looks very beautiful.   Before I began
to study botany, I had never noticed the blossoms
of the forest trees, and when I was shown the

light-feathery flowers of the Lime, and the
gracefully-drooping ones of the Sycamore, I
was quite astonished.   The flowers of the Sycamore grow in a drooping raceme; the calyx is
divided into five parts, but as it is scarcely
distinguishable from the petals, which are five
in number, and placed alternately with the
sepals, it appears to be in ten divisions (see *a*).
These flowers are partly male and female (see
*b* and *c*), and partly perfect.   In the perfect
flowers there are eight stamens, and two stigmas; and the ovary when ripe expands into a
curiously winged pod, called a samara (*d*), but
differently shaped to the samara of the Ash,
the thickened parts at the base of which contain
the seeds.   There is no albumen in the seed,
which, when put into the ground, expands into
two long thin cotyledons, (*a* in *fig*. 124 ) which,
if once pointed out, will always be known again
instantly.   If a ripe seed be opened when quite
fresh, the cotyledons or seed leaves will be found
within it, fresh, green, and succulent ; and these
leaves (*a* in *fig*. 124), which rise above the
ground as soon as the seed begins to germinate,
differ widely in shape from the true leaves (*b*)
which are serrated, and of a much thicker
texture.   The bracts of the Sycamore (*f* in *fig*.
123) are thick and leathery, and of a rich dark
brown.   The leaves are serrated at the margin ;

and the lower ones are cut into five lobes ; but
those near the flowers have generally only three

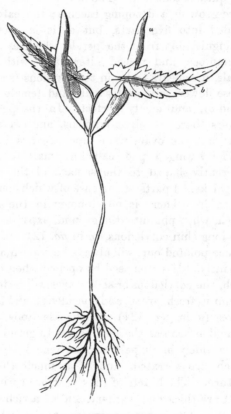

FIG. 124.—YOUNG SYCAMORE.

lobes (*e*), and in all the leaves, two of the lobes
are not so deeply cut as the others.

There are many species of Acer, most of
which are tall trees ; and they are chiefly distinguished from each other by the shape of the
leaves and of the samaras, or keys, the wings of
which, in some species, are near together, as
shown at *d* in *fig*. 123, and in others widely
apart, as in the common hedge Maple (*A. cam-
pestris*), and in the Norway Maples, as shown at
*a* in *fig*. 125.   This figure represents the flowers
of the Norway Maple (*Acer platanoides*), which
are in what botanists call a corymb, and stand
erect, instead of drooping like those of the
sycamore.   The leaves are deeply five-lobed,
and the lobes are so coarsely toothed, that the
teeth have almost the appearance of lobules.
The buds of this plant in winter are large and
red, and when they open in spring, the bracts
(*b*) curl back over the scales (*c*).   The leaves
become of a clear yellowish red in autumn, and
the whole plant is very ornamental.   When a
leaf of this tree is broken off, a milky sap issues
from the broken petiole or leaf-stalk, which is
of an acrid nature ; differing in this respect,
materially from the sap of the trunk, which is
very sweet.   Sugar indeed may be made from the
sap of the trunk of almost all the Maples ; but
particularly in America, from that of the Sugar

Maple (*Acer saccharinum*).   The flowers of the
red American Maple (*Acer rubrum*) are red, and
as from their colour, and their appearing a
fortnight before the leaves, they are very conspicuous, I have given a magnified representation of them in *fig.* 126, that my readers may
have an opportunity of examining the male

Fig. 125.—Flowers and Samara of the Norway Maple.

and female flowers from a living tree. In *fig.*
126, *a a* are male flowers, having no stigmas;
and *b b* are female ones, having no stamens.

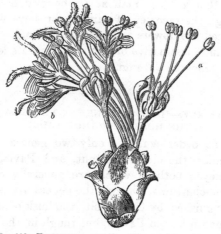

FIG. 126.—FLOWERS OF THE RED MAPLE (*Acer rubrum*).

The leaves of *Acer rubrum* become red in
autumn. The Tatarian Maple differs from the
other species in having entire leaves, and the
samaras are red when young; but all the other
kinds of Acer common in British gardens bear
a strong family likeness to each other. The
Ash-leaved Maple is now made into a separate
genus, and is called *Negundo fraxinifolia*. This
tree is easily distinguished from the Maples by
its compound leaves, which resemble those of

Y

the Ash, and its long pea-green shoots, which
have very few buds. The male and female
flowers of the Negundo are on different trees,
and they are so small as to be seldom seen,
though the racemes of samaras or keys which
succeed the flowers are very conspicuous. The
Negundo is a native of America, and its leaves
turn yellow in autumn.

---

### ORDER XLV.—HIPPOCASTANEÆ, OR ÆSCULACEÆ.
### THE HORSE-CHESTNUT TRIBE.

THIS order contains only two genera; viz.,
Æsculus, the Horse-chestnut, and Pavia, the
Buckeye; both of which are generally called
Horse-chestnuts, though the genera are easily
distinguished by their fruit, the husk of which
is smooth in the Pavias, but rough in the true
Horse-chestnuts. The buds of all the species of
both genera are covered with bracted scales,
most of which fall off when the leaves and
flowers expand; and those of the common Horse-
chestnut (*Æsculus Hippocastanum*) are very large,
and covered with a kind of gum. Four large
compound leaves, each consisting of five or seven
leaflets, and a raceme of sixty-eight flowers, have
been unfolded on dissecting one of these buds,
before the leaves unfold in spring. The flowers
of this species are produced in large, upright

panicled racemes (see *a* in *fig.* 127) ; and the
leaves (*b*) are compound, consisting of five or

FIG. 127.—FLOWERS OF THE HORSE-CHESTNUT.

seven leaflets, disposed in a palmate manner.
Two of the inner bracts, which remain after the
outer scales (which are very numerous) have
fallen, are shown at *c.* I mention this particularly, as these remaining bracts have very
much the appearance of stipules, and it is one
of the characters of the Horse-chestnuts that
their leaves are without stipules. The flowers
consist of five petals, two of which (*d* in *fig.*
128) are somewhat smaller than the others.
Each petal consists of a broad blade or limb

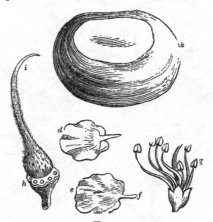

FIG. 128.—HORSE-CHESTNUT.

(*e*), and a very narrow claw (*f*). There are
seven stamens, three of which (*g*) are shorter
than the others. The filaments are inserted in

the receptacle (*h*), and surround the pistil, which is hairy, and has a long style and a curved stigma (*i*). The ovary is two-celled, and each cell contains two ovules, but seldom more than one seed ripens. The nut (*k*) is large, and covered with a shining brown skin, which is strongly marked with the hilum. When put into the ground, the cotyledons do not appear in the shape of seed-leaves, but remain in the ground, and the plumule and radicle are protruded as shown in *fig.* 129. The Acorn germinates in a similar manner, as already shown in *fig.* 86 in p. 192.

FIG. 129.—YOUNG PLANT OF HORSE-CHESTNUT.

The flowers of the different species of Æsculus vary considerably; as, for example, in the Scarlet Horse-chestnut (*Æ. rubicunda*), the calyx is tubular (see *a* in *fig.* 130), and there are but four petals, the upper two of which (*b*) are narrower than the lower ones (*c*), and have bearded

claws. This species has sometimes eight stamens. In the Yellow Horse-chestnut, or yellow
flowered Ameri-
can Buckeye,
the upper petals
(*a* in *fig*. 131)
are very much
smaller than the
lower ones (*b*),
and both have
very long claws.
There are four
petals, which
conceal the sta-
mens, of which
there are fre-
quently only six.
The seed of Pa-
via has only a
small hilum,
which resembles
the pupil of an
eye (see *fig.*
132); and hence
the genus has re-
ceived its Ame-
rican name of
Buckeye. In
one species (*P.*

FIG. 130.—SCARLET HORSE-CHESTNUT.

FIG. 131.—YELLOW HORSE-CHESTNUT.

*macrostachya*), the nut is eatable, and very much resembles that of a Sweet Chestnut when

FIG. 132.—NUT OF THE BUCKEYE.

boiled in milk. The stamens in this species are much longer than the petals, and they give a peculiarly light and elegant appearance to the flowers ; which, unlike those of the other species, do not appear till the latter end of summer or autumn.

## ORDER XLVI.—RHIZOBOLEÆ.—THE CARYOCAR TRIBE.

TREES of large size, natives of tropical America. *Caryocar nuciferum* produces the Suwarrow, or Butter-nut of the fruiterers' shops.

## ORDER XLVII.—SAPINDACEÆ.—SOAP-TREE TRIBE.

THE only plant in this order which will grow in the open air in England is *Kölreuteria paniculata,* a beautiful tree, with very elegant leaves, and panicles of yellow flowers, which are succeeded by a bladdery capsule, which is divided into three cells in its lower part, though it is only one-celled above. The rind and pulp of the fruit of *Sapindus Saponaria* are used as

soap in those countries of which it is a native.
The nuts of this plant are round and hard, and
of such a shining black that they are made
into buttons and beads by the inhabitants of
Spanish America. The whole plant, if thrown
into ponds containing fish, will intoxicate, and
sometimes kill them. Another interesting plant
belonging to this order is the Chinese fruit
called Litchi (*Euphoria* or *Nephelium Litchi*);
which has its sweet eatable pulp enclosed in a
kind of nut, much wrinkled oń the outside; so
that the fruit lies within the stone, instead of
being on the outside of it. These hard, stone-
like berries grow in loose racemes.

---

### ORDER XLVIII.—MELIACEÆ.—THE BEAD-TREE TRIBE.

*Melia Azederach*, the Pride of India, or Indian
Lilac, or Bead-tree, for it is known by all these
names, is a native of Syria, which has become
almost naturalised in the South of Europe,
particularly near the Mediterranean. The
leaves are bi-pinnate, the flowers are violet-
coloured, and the fruit, which resembles that of
the cherry, is of a pale yellow when ripe. The
pulp is poisonous, and the stones are used for
making rosaries in the Roman Catholic countries.

## ORDER XLVIII*.—CEDRELEÆ.—THE MAHOGANY TRIBE.

This order was at first united to Meliaceæ by De Candolle, but it has been separated on account of its winged seeds. It contains, among other genera, the Mahogany tree (*Swietenia Mahagoni*), and the West Indian Cedar (*Cedrela*). The leaves of these trees are alternate and pinnate, with unequal-sided leaflets ; and the flowers are in large spreading panicles composed of numerous little cymes. The fruit is capsular, and the seeds are winged. The genera contained in this order, all require a stove in Great Britain.

## ORDER XLIX.—AMPELIDEÆ.—THE VINE TRIBE.

The natural order Ampelideæ contains several genera, but of these only the Vine and the five-leaved Ivy are common in British gardens. It seems almost ridiculous to talk of the flowers of the Vine, as the bunches, even when they first appear, seem to consist of only very small grapes, which gradually become large ones. The flowers, however, though small and insignificant, are perfect, and they have each a distinct and regularly formed calyx and corolla. The calyx of the common Grape (*Vitis vinifera*) is very small, and remains on till the fruit is ripe ; there are five petals (*a* in *fig.* 133), which never expand, but remain fastened together at the tip, detaching themselves at the base, when

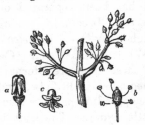

it is necessary that they should give room to the ripening sta- mens (*b*). The petals, which form a kind of extinguisher, when they are raised by the five stamens, fall off

FIG. 133.—FLOWERS OF THE VINE.

(*c*), and occasion the chaffy appearance observable in clusters of Vine- flowers. The ovary is, when young, in two cells, each containing two seeds ; and it is crowned with a nearly flat, round stigma, with- out any style. When the fruit begins to swell, the ovary becomes filled with a pulp, which is solid, and not contained in bags like that of the Orange ; and the dissepiment that divided the two cells gradually wastes away. Two, and sometimes three of the seeds also frequently disappear, so that four seeds are rarely found in the ripe grape. The seeds themselves are bony, and covered with a jelly-like matter ; and when they are cut open, they are found to consist of a large quantity of hard albumen, with a very small embryo at the tip. The Vine is a climbing shrub, with lobed leaves, which

are frequently deeply cut ; the bunches in which
the grapes are disposed are called branched or
thyrsoid racemes (see 137), and the tendrils,
by which the plant climbs, are supposed to be
abortive peduncles, drawn out into these long,
flexible, curling bodies, instead of producing
bunches of grapes. The footstalks of the leaves
are articulated, and will separate from the
branch without tearing them.　The different
species of vines differ from each other chiefly
in their leaves ; but in the American grapes
the calyx is sometimes entire, and sometimes
the stamens and pistils are in different flowers.

The five-leaved Ivy, or Virginian Creeper
(*Ampelopsis hederacea*), differs very little from the
Vine in the botanical construction of its flowers.
The calyx is, however, almost entire, and the
five petals separate in the same way as those
of other flowers; but in other respects they
closely resemble those of the Vines.　The berries
are small, and not palatable, though they might
be eaten with perfect safety.　The leaves are
palmate, and they are divided into three or five
stalked leaflets.　The stems are climbing and
rooting; and the leaves take a beautiful deep
red in autumn.　The genus Cissus also belongs
to this order.

ORDER L.—GERANIACEÆ.—THE GERANIUM TRIBE.

THE order Geraniaceæ contains several genera of well-known plants, the most popular of
which are *Pelargonium, Erodium,* and *Geranium,*
signifying Stork's-bill, Heron's-bill, and Crane's-
bill, which differ very slightly from each other.
The greenhouse Geraniums, which are all either
natives of the Cape of Good Hope, or hybrids
raised in Europe from the species originally
imported, were, till lately, all included in the
genus *Pelargonium;* but what were sections of
that genus have, by some botanists, been now
made separate genera. As probably, however,
this rage for giving new and different names to
divisions and subdivisions will not be generally
adopted, I will not trouble my readers with
any other distinctions than those between the
three leading genera; and even these, I think
they will allow, appear very trifling. The calyx
of the Pelargonium is in five sepals, and two of
them end in a kind of spur; which is, however,
not very perceptible, as it runs down the peduncle or footstalk of the flower, and grows to it,
so as to seem only a part accidentally enlarged.
The corolla is in five petals, the upper two of
which are generally larger, and differently
marked to the others. Sometimes there are
only four, and sometimes there are six petals;

but these are exceptions to the general rule.
The perfect stamens vary in number from four
to seven; but there are always ten filaments,
which are dilated, and grow together at the
base; and I was quite delighted with the
sparkling gem-like appearance of the membrane which they form when thus united, when
I looked at it through my little microscope.
In the plant now before me (a hybrid called the
Duke of Sussex), the upper parts of some of the
stamens have turned into little petals, retaining
the white membrane-like part at the base, and
thus curiously exemplifying the manner in which
double flowers are formed, which is always by
the metamorphosis of the stamens, or of the
stamens and pistil, into petals. The pistil of
the Pelargonium appears, when young, to consist of a five-celled ovary, with a long slender
style, the tip of which is divided into five slender
curved stigmas. The cells of the ovary are,
however, five one-seeded carpels, each having a
separate style; and though both the carpels
and styles appear firmly grown together when
young, yet, in fact, they only adhere to an
elongation of the receptacle (see *a* in *fig.* 134),
which is here called the central axis, and from
which, when ripe, they part with elasticity, and
curl up, as shown at *b;* the styles, or awns, as
they are sometimes called, being hairy inside.

The shape of the unripe seed-vessel, with its persistent calyx, is shown at *c*, and a detached

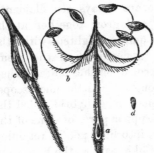

seed at *d*. No plant hybridises more freely than the Pelargonium; and thus, the number of new kinds raised every year defies all description, and they have been so mixed and intermixed with each other, that it is

FIG. 134.—SEED-VESSEL OF A PELARGO-NIUM.

not easy to say to what species the most splendid hybrids are allied. A few species, however, remain nearly unchanged, and the best known of these are *P. zonale*, the Horseshoe Geranium; *P. inquinans*, the common scarlet, the juice of the leaves of which is said to stain the fingers brown; *P. graveolens*, and *P. capitatum*, the rose-scented Geraniums, and *P. tricolor*. All the Pelargoniums have their flowers in heads or umbels; and the calyx in all of them remains on till the seeds are ripe. The seed-vessel, or fruit, as it is called by botanists, is long and pointed, forming some resemblance to the head of a stork; the ovary shrouded in the persistent calyx, representing the head of the bird, and the long styles the beak. The leaves vary in shape in the different kinds: sometimes they are roundish, as in the Horseshoe Geranium, and marked with a dark band or zone, whence the specific name *zonale*; and sometimes they are deeply cut, as in the rose-scented kinds: some are shrubby, and some herbaceous; and the stems of some species are warted, and the roots of others tuberous.

The genus Erodium consists principally of European plants, three of which are natives of England. The commonest of these (*Erodium cicutarium*) is called in many parts of England the Wild Geranium; and nearly allied to it, but less common, is *E. moschatum*. The principal points in which this genus differs from Pelargonium are, that the filaments of the stamens are very little united at the base; that there are always five filaments which bear anthers, and five that are sterile, and that the latter have each a gland at the base. The calyx is also without the spur, and the seed-pod is thought to resemble a heron's head more than that of a stork. When it bursts, also, the styles, which are hairy inside like those of the Pelargonium, do not curl up in the same manner as in that genus, but spirally.

The genus Geranium differs from Erodium principally in having the stamens all perfect;

but the alternate ones are longer than the others, and have a gland at the base of each. The seed-pod is said to resemble the head of a crane, and when it bursts, the styles, which are smooth inside, curl up round and round like the coil of a rope. The seeds of many of the kinds are beautifully netted. Many of the species are British weeds, and among the commonest of these may be mentioned Herb Robert (*Geranium Robertianum*), and the Meadow Crane's-bill (*G. pratense*). Dr. Lindley, in his Ladies' Botany, mentions a curious and beautiful experiment which may be performed by pressing the petals of a Geranium between two pieces of glass which have been previously wetted. He says, that by pressing the two glasses firmly together, all the air may be squeezed out of the petal, and it will become transparent. " You may then," he adds, " with a pretty good magnifying power, observe all the air-vessels of the veins distinctly, looking like fine threads of silver-wire twisted up like a spiral spring. It is on account of this appearance that the air-vessels are called, technically, spiral vessels." The experiment appeared to me so easy, and at the same time so interesting, that I tried it, but unfortunately without success; probably owing to the want of power in my microscope.

### ORDER LI*.—LIMNANTHEÆ.

THIS little order contains only one plant, *Limnanthes Douglassi*, a pretty Californian annual, with yellow and white flowers. It resembles Geraniaceæ in its botanical construction, but it does not discharge its seeds with elasticity.

---

### ORDER LI.—TROPÆOLACEÆ.—THE NASTURTIUM TRIBE.

THE well-known flowers called Nasturtium, or Indian Cress, give their name to this order; which, in fact, consists only of the genus Tropæolum, and an obscure genus not yet introduced. In the flowers of the Nasturtium, the calyx and corolla are of nearly the same colour, but they may be easily distinguished from each other. The calyx is drawn out into a spur behind, and the petals, which are unguiculate, or claw-shaped, are fringed at the base. The leaves and stem are succulent, and have the taste of cress, and hence the plant has received its popular name,—Nasturtium being the botanic name of the water-cress. The Tropæolum has five petals, eight stamens, and three carpels, which are joined together into a trigonal fruit, each carpel containing one seed, which adheres to it. The embryo is large, and fills the whole

z

seed, which is without albumen. The unripe
carpels are sometimes pickled, and used as a
substitute for capers. The whole plant has not
only the taste, but the properties, of cruciferous
plants ; and even the caterpillars of the cab-
bage-butterflies feed upon it.

-----

## ORDER LII.—BALSAMINEÆ.—BALSAM TRIBE.

NEARLY allied to the Geraniums, and resembling them, in the opening of the seed-pods, are
the beautiful plants contained in the order Bal-
samineæ. The two genera best known in British gardens are Balsamina and Impatiens.
The common Balsam (*Balsamina hortensis*), has
a small green calyx of two sepals; there are
four petals, one of which is drawn out into a
short spur at the base. There are five stamens,
each bearing a two-celled anther. The ovary
is one-celled, but it separates into five valves,
when the seeds are ripe, bursting with elasticity, and the valves curling inwardly from the
apex to the base. There are five stigmas, quite
distinct from each other, and appearing just
above the ovary, without any style; and the
peduncles are simple and one-flowered.

The genus Impatiens, which contains the
common Noli-me-tangere, or Touch-me-Not,
and other similar plants, though it agrees with

Balsamina in having five anthers, has only three
of them with two cells, the others having one
cell each.   The stigmas also are joined together at the base, and the capsule bursts at the
slightest touch, the valves coiling up spirally
from the base to the apex, and detaching themselves from the plant at the same time that they
expel the seeds.   The peduncles grow from
the axils of the leaves, and they are branched
and many-flowered.   A separate order, called
Hydrocereæ, has been made of one of the species
of Impatiens (*I. natans*).   It is an aquatic plant,
a native of the East Indies.

ORDER LIII.—OXALIDEÆ.—THE WOOD-SORREL
TRIBE.

THE flowers of all the species of Oxalis, the
Wood-sorrel, are very pretty.   The flowers
have five regular petals, each furnished with a
claw ; and the petals are spirally twisted in the
bud.   There are ten stamens, and five styles.
The capsule is five-celled, and five or ten valved,
the valves opening lengthways.   Most of the
species are natives of South America, and
greenhouse plants in England.

ORDER LIV.—ZYGOPHYLLEÆ.—THE BEAN-CAPER
TRIBE.

THE flowers of the Bean-caper are usually
yellow; and the five petals are long, narrow,
and placed widely apart. The botanic name
of Zygophyllum signifies " with the leaves in
pairs," and this is the case to a remarkable
degree. *Fagonia cretica* is a very pretty plant,
with purple flowers very much like those of
Clarkia ; and *Guiacum*, the Lignum Vitæ, is
remarkable for the hardness of its wood and
the gum it produces. Melianthus belongs to
this order.

ORDER LV.—RUTACEÆ.—THE RUE TRIBE.

THIS order has been divided into four sections ; three of which contain well-known plants,
and have been divided into three orders by
many botanists. The Rue (*Ruta graveolens*) is
well known from its strong and disagreeable
smell, which is produced by the oil secreted in
transparent cells in the leaves, which have the
appearance of dots, when the leaves are held up
to the light. The leaves are of a bluish green,
and the flowers of a greenish yellow ; the latter
growing in cymes at the end of the branches.
There are four sepals, four petals, snd eight stamens. There are four carpels, seated on an elevated receptacle, and each containing one cell,

which grow into a four-celled fruit. In Fraxinella
(*Dictamnus*) the petals are unequal; there are
ten stamens, one style, and the carpels are
two-seeded.   In Diosma there are only five
stamens, the style is arched, and the capsule
consists of five-horned carpels.   In Corræa the
leaves are opposite; there are eight stamens,
and the four petals grow together into a tube
at the base; and in Crowea there are five
sepals, five petals, and ten stamens; the leaves
are also alternate.   The Diosmas have as strong
a scent as the Rue, and a perfume is made from
them called Bucku at the Cape of Good Hope,
of which country they are natives.

The section Zanthoxyleæ contains the Zanthoxylum, also called the Toothache Tree, or
Prickly Ash, a native of North America, the
bark of which is very fragrant, and is said to be
a cure for toothache and rheumatism; Ptelea
or Shrubby Trefoil; and *Ailantus glandulosa*.
*Zanthoxylum fraxineum* has very pretty pinnate
leaves, and small purple flowers; *Ptelea trifo-
liata* has curiously winged fruit, which resemble
those of the elm; and the Ailantus has remarkably long compound leaves, one leaf having
been known to have fourteen pairs of leaflets,
and to be upwards of three feet long.   The two
following orders are included in Rutaceæ by
some botanists.

## ORDER LVI.—SIMARUBACEÆ.

*Quassia amara,* the bark of which is sometimes used as a substitute for hops, is perhaps the best known plant belonging to this order. All the species are trees or shrubs, natives of tropical America, with bitter bark, milky juice, and pinnated leaves.

## ORDER LVII.—OCHNACEÆ.

Tropical shrubs with yellow flowers and shining leaves; seldom seen in British hothouses.

## ORDER LVIII.—CORIAREÆ.

Only one species of this order is common in British gardens, viz. *Coriaria myrtifolia;* the leaves of which are astringent, and used in dyeing black, and the berries are poisonous.

### § II.—Calycifloræ.

The plants comprised in this division have their petals and stamens inserted in the calyx, or in a lining of it formed by the dilated receptacle.

## ORDER LIX.—CELASTRINEÆ.

THIS order is divided into three sections, each
containing well-known plants. The first of these
takes its name from *Staphylæa pinnata,* the
Bladder-nut. In the flowers of this plant the
calyx is in five divisions, and white tinged with
pink, so as to be scarcely distinguishable from
the corolla. There are two or three carpels,
which are surrounded by the receptacle, and
the styles of which adhere slightly together.
The capsule is bladdery, and consists of two or
three cells, each containing one smooth, brownish, bony seed, which looks as though one end
had been cut off at the hilum. The leaves are
compound, each having five leaflets. The second
section contains, among other plants, the
Spindle-tree (*Euonymus europæus*), Cassine,
and the Staff tree (*Celastrus scandens*). The
Euonymus has small whitish-green inconspicuous flowers; but it is remarkable for the
beauty of its capsules, which are fleshy, and of
a bright rose-colour, while the seeds, which are
of a bright orange, are enwrapped in a covering
called an aril, by which they remain attached
to the capsule after the valves have opened.
Each capsule has five cells and five seeds, and
each seed has a little white stalk attached to its
aril, like the funicle of a pea. There are several

species. The Celastrus is a climbing shrub, remarkable for its clusters of flowers, but which has nothing else to recommend it. The third section, Aquifoliaceæ, is made a separate order, under the name of Ilicineæ, or Aquifoliaceæ, by many botanists; some of whom place it in the sub-class Corollæfloræ, because the petals are connected at the base. The most common plants that it contains are included in the genera Ilex and Prinos. In *Ilex aquifolium*, the Holly, the corolla (*a* in *fig*. 135), is in four or five petals

FIG. 135.—THE HOLLY.

connected at the base; there are four stamens, the cells of the anthers of which adhere to the sides of the filament (*b*). The berry (*c*) is four-

celled, each cell containing a one-seeded nut.
The leaves (*d*) are simple, and smooth, shining
and prickly at the edges, which are curved
upwards.   Prinos, the Winter-berry, is a little
evergreen shrub, with red berries.

---

## ORDER LX.—RHAMNACEÆ.

THE most interesting genera in this order are
Paliurus, Zizyphus, Rhamnus, and Ceanothus.
Christ's Thorn (*Paliurus aculeatus*) is easily
known by its crooked prickly stem, and its singular fruit, which, from its resembling a head
with a broad flat hat on, the French call,
Porte-chapeau.   The flowers are yellow, but
they are too small to be ornamental.  *Zizyphus
Jujuba* differs from Paliurus chiefly in its fruit,
which resembles a small plum, and from the
fruit of which the Jujube lozenges are made.
There are numerous species of Rhamnus, some
of which are trailing-shrubs, and others low
trees.   Some of the species, such as *R. Alater-
nus*, are evergreen shrubs, very useful in town-
gardens, as they are not injured by smoke ;
others, such as the Purging Buckthorn (*R. ca-
tharticus*), have deciduous, rough, feather-nerved
leaves, and the branchlets terminating in a
thorn.  The berries of the plants in this division
are sold for dyeing yellow, under the name of

French or Avignon berries. Another division
includes the species which are without thorns.
All these plants have their male and female
flowers distinct. The last division of Rhamnus
has perfect flowers, and dark-purple berries, as
for example, the Berry-bearing Alder (*R. fran-
gula*). The genus Ceanothus is well known
from the beautiful *C. azureus*. The other species
have generally the same kind of terminal,
upright panicles of feathery flowers, but they
are very inferior in beauty. *C. americanus*, which
has white flowers, is sometimes called American
Red-root, or New Jersey Tea.

### ORDER LXI.—BRUNIACEÆ.

SMALL heath-like shrubs, natives of the Cape
of Good Hope.

### ORDER LXII.—SAMYDEÆ.

TROPICAL shrubs or trees with dotted leaves,
and inconspicuous flowers.

### ORDER LXIII.—HOMALINEÆ.

THIS order contains the handsome evergreen
half-hardy shrub, *Aristotelia Macqui;* the
flowers are insignificant, but the berries are

black, acid, and eatable, and the leaves are
smooth, shining, and so abundant as to render
the plant an excellent screen.

---

### ORDER LXIV.—CHAILLETIACEÆ.

AFRICAN plants, with panicles of small white
flowers, and simple leaves.

---

### ORDER LXV.—AQUILARINEÆ.

TREES, natives of Asia, little known in England.

---

### ORDER LXVI.—TEREBINTHACEÆ.—THE
TURPENTINE TRIBE.

THIS order is divided by De Candolle into
seven sections; viz., 1. Anacardiaceæ, including the Cashew-nut (*Anacardium*), the Mango
(*Mangifera*), and the Turpentine trees (*Pistacia*); 2. Sumachineæ, which contains *Rhus,
Schinus*, and *Duvaua*: 3, Spondiaceæ, containing the Hog-plum (*Spondias*); 4. Burseraceæ,
including the Jamaica Birch (*Bursera*), and
the Balm of Gilead tree (*Balsamodendron*);
5. Amyrideæ, the West Indian Balsam tree
(*Amyris*); 6. *Spatheliaceæ*, the West Indian
Sumach (*Spathelia*); and 7. Connaraceæ,

containing Omphalobium, and other exotic genera. Of these modern botanists make five distinct orders, viz., Anacardiaceæ, including the first, second, and fifth sections ; Amyrideæ, Spondiaceæ, Burseraceæ, and Connaraceæ. Ptelea, which was originally included in this order, is now generally placed in Xanthoxylaceæ.

The plants contained in this order have in some cases perfect flowers, and in others, the male and female flowers on different plants. They all abound in a resinous gum ; that from the Mastic tree (*Pistacia Lentiscum*), and several of the species of Rhus, is used for making varnish ; the gum of the Turpentine tree (*P. Terebinthus*) is the Chian or Cyprus turpentine. The flowers are small, and generally produced in panicles, the petals are sometimes wanting. The leaves are alternate, without stipules, and often compound. The flowers have generally five petals, and five or ten stamens ; and the fruit is drupaceous, or capsular, varying in the different genera. In Anacardium, the peduncle which supports the Cashew-nut is fleshy and pear-shaped, so as to resemble a fruit more than the nut itself. The Mango has a fleshy drupe, with a woody, fibrous stone or nut. In Pistacia, the fruit is a dry drupe inclosing a nut, which is eatable in *P. vera.* Both the male and female flowers in this genus

are handsome, though without petals, from the
anthers being yellow, and the stigmas crimson.
The different species of Sumach, or Rhus, are all
poisonous ; and the Venetian Sumach (*Rhus
cotinus*) is remarkable from the appearance
presented by its flower-stalks in autumn ; as all
the flower-stalks which do not bear fruit dilate,
after the flowers have dropped, and become
covered with a great quantity of white cottony
hair, which makes each panicle resemble a
powdered wig ; and hence, the French call the
tree *Arbre à perruque.*

---

ORDER LXVII.—LEGUMINOSÆ.—(See Chap. II.
in P. 35.)

The plants belonging to this order have alternate leaves, which are generally compound, and
frequently have the common petiole tumid ;
they have also two stipules at the base of the
petiole, and frequently two others to each
leaflet. The pedicels are usually articulated,
and the flowers are furnished with small bracts.
The flowers have a five-parted calyx, and a
corolla, sometimes papilionaceous, and sometimes spreading, which has never more than five
petals, though it has frequently less. The fruit
is a legume, though sometimes, when there is
only one seed, it has the appearance of a drupe.

There are eleven sections given in De Candolle's
Prodromus, viz., 1. *Sophoreæ*, the Sophora tribe ;
2. *Loteæ*, the Lotus tribe; 3. *Hedysareæ*, the
Sainfoin tribe ; 4. *Vicieæ*, the Vetch tribe,
(including the Pea and Bean) ; 5. *Phaseoleæ*,
the Kidney-bean tribe ; 6. *Dalbergia*, the Gum-
dragon tribe; 7. *Swartzia* ; 8, *Mimoseæ*, the
Mimosa tribe ; 9. *Geoffrea*, the Earth-nut tribe,
(including the Earth-nut *Arachis*, and the Ton-
quin Bean, *Dipterix*) ; 10. *Cassieæ*, the Cassia
tribe ; and 11. *Detarieæ*. Some botanists include Moringa, the Horse-radish tree, in Leguminosæ, but others make it a separate order
under the name of Moringeæ.

---

ORDER LXVIII.—ROSACEÆ.—(See Chap. III. in P. 50.)

The flowers have five sepals, combined in
their lower part into a tube, but divided above
into five lobes ; and the corolla has generally
five petals. There are numerous carpels, which
are usually inclosed in the fleshy tube of the
calyx. The ovary is one-celled, and there is
seldom more than one seed, and scarcely ever
more than two. The leaves are alternate,
generally compound, and always furnished with
stipules. De Candolle divides the order into
eight tribes, viz., 1. *Chrysobalaneæ ; 2. Amygda-*

*lineæ* ; 3. *Spiraceæ* ; 4. *Neuradeæ* ; 5. *Dryadæ*, or *Potentilleæ* ; 6. *Sanguisorbeæ* ; 7. *Roseæ* ; and 8. *Pomaceæ*. Of these, the first, second, third, and eight, are made separate orders ; the fifth, sixth, and seventh are retained in Rosaceæ. Neuradeæ was first removed to Ficoideaceæ, and afterwards made a separate order ; and another order has been made, called Quillageæ, including only the genera Kageneckia and Quillaja.

## ORDER LXIX.—CALYCANTHACEÆ.

THERE are only two genera in this order, both of which are remarkable for the fragrance of their flowers. The American Allspice (*Calycanthus floridus*) is a shrub, with very dark blackish purple flowers, which botanists consider to be all calyx, the plants in this order having no petals. The lobes of the calyx are somewhat leathery in texture, and lanceolate in form ; they are very numerous, and they are disposed in several rows, like scales. The stamens are numerous, but only the outer twelve are fertile, and they soon fall off. The peduncle is thickened below the flower ; and the receptacle is dilated, and drawn out over the carpels, which are arranged in it like those of the rose, which they closely resemble, but are much

larger. The leaves are opposite and feather-
nerved. *Chimonanthus fragrans*, so well-known
for its beautiful yellowish flowers, which are
produced about Christmas, belongs to this
order. In this plant the lobes of the calyx are
oval, and not nearly so numerous as in Calycanthus; the outer lobes look like bracts. The
stamens are less numerous, and not deciduous;
and only five are fertile, which are united at the
base. This plant was formerly called *Calycanthus præcox*.

### ORDER LXX.—GRANATEÆ.

This order has only one genus and two
species. The Pomegranate (*Punica Granatum*)
has a tubular calyx, with a limb in five or
seven divisions, and the same number of petals
as there are segments to the calyx. The calyx
and corolla are both of the same colour. When
the petals fall, the tube of the calyx swells, and
becomes a many-celled berry, the limb of the
calyx remaining on, and forming a kind of
crown to the fruit. The cells are divided into
two parts, and they contain a great number of
seeds which are plunged in a juicy pulp. The
other species, *P. nana*, only differs in being a
dwarf plant, and in the leaves being narrower.
The Pomegranate was formerly included in
Myrtaceæ.

### ORDER LXXI.—MEMECYLEÆ.

TROPICAL trees and shrubs, with white or purplish flowers, and eatable fruit.

---

### ORDER LXXII.—COMBRETACEÆ

THIS order is well-known from the two beautiful climbing stove-plants, *Combretum purpureum*, and *Quisqualis indica*. The flowers of the former are disposed in racemes, which have a peculiarly light and graceful appearance, from the great length of their stamens; and as they are of a brilliant scarlet, the name of Purpureum is very ill applied to the species. The flowers of *Quisqualis indica* have a very long slender tube to the calyx, and five velvet-like petals, which vary in colour from a yellowish white to red, changing in the course of one day.

---

### ORDER LXXIII.—VOCHYSIEÆ.

BRAZILIAN trees and shrubs, with yellow flowers, and stipulate, feather-nerved leaves.

---

### ORDER LXXIV.—RHIZOPHOREÆ.

THE Mangroves (*Rhizophora*) are tropical trees, growing in the soft mud of rivers, particu-

A A

larly in that of the Niger, so that, when the
rivers are full, they appear to grow out of the
water.   The seeds have the singular property
of germinating in the capsule, and sending down
long roots while yet hanging on the tree, the
branches of which thus appear, at a little distance, as if covered with long white strings.
All the genera belonging to this order require
a stove in England.

## ORDER LXXV.—LOPHIREÆ.

THE only plant contained in this order is a
beautiful shrub from Sierra Leone, with terminal corymbs of white flowers, and coriaceous
leaves.

## ORDER LXXVI.—ONAGRARIÆ.—(SEE CHAP. IV.
IN P. 75.)

THE tube of the calyx generally adheres to
the ovary, and its limb is usually two or four
lobed, the lobes frequently adhering together.
The petals are either four, or equal in number
to the lobes of the calyx ; they are inserted
in the mouth of the tube, and are twisted in the
bud.   The fruit is generally a capsule, or a
berry, with two or four cells ; and there are
numerous seeds.   The leaves vary consider-

ably, and are sometimes alternate, and sometimes opposite, but never compound. De Candolle divides this order into six sections: viz. 1. *Montinieæ ;* 2. *Fuchsieæ ;* 3. *Onagreæ,* containing the Evening Primrose (*Œnothera*), and the French Willow Herb (*Epilobium*) ; 4. *Jussieuæ :* 5. *Circææ,* including the Enchanter's Nightshade (*Circæa*), and *Lopezia ;* and 6. *Hydrocaryes,* containing the Water-caltrops (*Trapa natans*). This last section is sometimes made a separate order.

---

### ORDER LXXVII.—HALORAGEÆ, or CERCODIANÆ.

Most of the plants in this order are British weeds ; as for example, the Water Milfoil (*Myriophyllum*), Water Starwort (*Callitriche*), and Mare's-tail (*Hippuris*) ; but some are natives of North America, China, &c., and one genus has lately been discovered in Australia, which Dr. Lindley has named *Loudonia aurea,* and which is a large shrub, with corymbs of golden yellow flowers.

---

### ORDER LXXVIII.—CERATOPHYLLEÆ.

British weeds called Hornwort.

### ORDER LXXIX.—LYTHRARIEÆ, or SALICARIÆ.

THE principal plants in this order that are interesting to amateurs, are included in the genera Lythrum, Cuphea, Heimia, Lawsonia, and Lagerstrœmia. The genus Lythrum contains all those showy British plants which are called the Willow Herbs. The flowers are purple, and the petals, which are four or six in number, are crumpled in the bud. The stamens are either the same number as the petals, or twice the number, and the capsule is two-celled. The calyx, as in all the plants included in this order, is tubular, with numerous lobes; and the petals soon fall off. Cuphea is a genus principally of annual plants, with six or seven dark purple petals, unequal in size, and curiously inserted in the calyx. Heimia is a genus of South American shrubs, with yellow flowers. Lawsonia inermis produces the Henna, which the ladies of the East use to dye the palms of their hands pink; and Lagerstrœmia is a beautiful conservatory tree, with handsome flowers. This plant is sometimes called the pride of India.

### ORDER LXXX.—TAMARISCINEÆ.—THE TAMARISK TRIBE.

THERE are very few plants in this order, and the only ones common in British gardens are

the French Tamarisk (*Tamarix Gallica*), and
the German Tamarisk (*Tamarix*, or *Myriacaria
Germanica*); both of which are easily recognized
by their light airy branches, (which when young
are covered with closely imbricated leaves,
though the leaves drop off as the wood ripens,)
and their terminal erect spikes of whitish or
pink flowers. The seeds are large, and are
each furnished with a tuft of hairs at the end of
a kind of stalk. These plants are very suitable
for planting near the sea, as they are uninjured
by the sea-breeze.

## ORDER LXXXI.—MELASTOMACEÆ.

THIS order consists of showy exotic plants,
most of which require a stove in Britain, and
which are easily known by their leaves being
marked with two or more deep lines running
parallel to the midrib. They are all free-grow-
ing plants, with very handsome flowers, which
are generally purple or white.

## ORDER LXXXII.—ALANGIEÆ.

THERE are two genera, Alangium and Mar-
lea, both handsome shrubs, natives of India.

## ORDER LXXXIII.—PHILADELPHEÆ.—THE MOCK-ORANGE TRIBE.

THERE are three genera in this order: viz., Philadelphus, the Mock Orange or Syringa; Decumaria and Deutzia, all which have white flowers. There are many species of Philadelphus, all of which are easily known by their large white flowers, and large coarse-looking leaves. The flowers of the common species (*P. coronarius*) smell like those of the Orange, and the leaves taste like cucumber. There is only one species of Decumaria (*D. barbara*), which is a native of Virginia and Carolina, and is a climbing shrub, with terminal corymbs of white, sweet-scented flowers. *Deutzia scabra*, though only introduced in 1833, is already common in gardens; and it is a general favourite from the great abundance of its flowers. Though it said to be not a true climber, its stems are too weak to stand without support. It is a native of Japan, and though generally kept in pots, it is supposed to be quite hardy.

## ORDER LXXXIV.—MYRTACEÆ.—THE MYRTLE TRIBE.

No plants are more easily recognized than those belonging to this tribe; as they are easily distinguished by their entire leaves, which have no stipules, and which, when held up to the

light, appear to be not only full of transparent
dots, but to have a transparent line round the
margin.   The flowers have also abundance of
stamens on long slender filaments which look
like tufts of silk, and only four or five petals.
The whole of the plants are fragrant, and every
part of them seems full of an aromatic oil, which
is particularly visible in the flower-buds of *Caryophyllus aromaticus*, which when dried form what
are commonly called cloves ; and in the leaves
of some of the kinds of Eucalyptus.   The genera may be divided into two sections, viz.,
those with a dry capsule for the fruit ; in which
are included *Melaleuca* and its allied genera,
*Eucalyptus, Callistemon, Metrosideros and Lepto-
spermum ;* and those with berry-like fruit, the
most interesting of which are *Psidium*, the Gu-
ava ; *Myrtus*, the Myrtle ; *Caryophyllus*, the
Clove ; *Eugenia or Myrtus pimenta*, Jamaica
Allspice ;   and *Jambosa Vulgaris or Eugenia
Jambosa*, the Rose Apple.   In some of the genera, as for example in Eucalyptus, the sepals
of the calyx become detached at the base, and
being united above form a sort of cap or calyptra, which is pushed off by the stamens when
the flower begins to expand.   Besides the plants
already enumerated, some botanists add another
section to Myrtaceæ, which others consider a
separate order, under the name of *Lecythideæ*.

This section contains three genera, the most remarkable plants in which are the Cannon Ball-tree (*Lecythis Ollaria*), and the Brazil Nut (*Bertholletia excelsa*). The fruit of this last plant is fleshy, and as large as a child's head, opening with a lid, and containing sixteen or twenty triangular seeds, laid over each other in a regular manner, which are the Brazil-nuts sold in the shops.

### ORDER LXXXV.—CUCURBITACEÆ.—THE GOURD TRIBE.

THE plants included in this order have generally the male and female flowers distinct. The calyx is tubular, and generally five-toothed; there are five petals usually connected at the base, and which have strongly marked reticulated veins. There are five stamens, four of which are united so as to form two pairs, with the fifth one free. The anthers are two-celled, and generally very long. There are three or five two-lobed stigmas, which are thick and velvety. The fruit is fleshy, with numerous flat seeds. The leaves are palmate, and very rough; and the plants have succulent stems, and climb by means of their tendrils. The principal genera are, Cucumis, which includes the Melon (*C. melo*), the Cucumber (*C. sativus*), the Mandrake

(*C. Dudaim*), the Water Melon (*C. citrullus*), and the Colocynth (*C. colocynthis*) ; Bryonia, best known by the White Bryony (*B. dioica*) ; Momordica, including the Balsam Apple (*M. balsamea*), and the Squirting Cucumber (*M. elaterium*) ; and Cucurbita, including all the kinds of Pumpkin (*C. pepo*), and Vegetable Marrow (*C. ovifera*). To these may be added Lagenaria, the Bottle Gourd ; and Trichosanthes, the Snake Gourd, plants far more curious than useful. Some botanists include the Papaw-tree (*Carica Papaya*) in Cucurbitaceæ, but others make it into a separate order under the name of Papayaceæ.

ORDER LXXXVI.—PASSIFLOREÆ.—THE PASSION-
FLOWER TRIBE.

THE plants belonging to this order may be instantly recognized by the very singular arrangement of the pistil and stamens. The receptacle is raised in the centre of the flower so as to form a long cylindrical stipe, on which is placed the ovary, with its three styles, each ending in a fleshy stigma; a little lower are five stamens, with their filaments growing together round the stipe, and with large anthers which are attached by the back. At the base of the stipe are two or more rows of filaments without

anthers, which are called the rays. There are
five petals and five sepals ; but some botanists
consider the whole to be sepals, and that the
petals are wanting. The fruit of some of the
species is eatable. It is about the size of a
large egg, and contains numerous seeds, which
are enveloped in a kind of pulp.

## ORDER LXXXVI.*—MALESHERBIACEÆ.

THIS order consists entirely of the plants belonging to the genus Malesherbia ; which are
mostly annuals, or biennials, with very showy
blue or white flowers, introduced from Chili in
1832. The genus was formerly included in
Passifloraceæ.

## ORDER LXXXVII.—LOASEÆ.

ALL the species contained in this genus are
natives of North America, and most of them
are annuals, with very showy flowers. The
genera *Loasa* and *Caiophora* are covered with
glandular hairs or bristles, which sting much
worse than those of the nettle. *Bartonia aurea*
is one of the most splendid annuals in cultivation, from its golden yellow flowers; *Blumen-
bachia* has the fruit roundish and spirally twisted,
and *Caiophora* has the fruit horn-shaped, and

twisted in a similar manner.   This curious construction of the fruit may be seen in *C. punicea*,
the well known showy climber, generally called
*Loasa aurantiaca*, or *lateritia.*   The fruit of the
true kinds of Loasa is plain and not twisted, as
may be seen in *L. nitida*, *L. Placei*, and in short
in all the other species of the genus.   The
flowers of most of the plants in this order are
very curiously constructed, there being two sets
of petals quite distinct in form and colour, and
two sets of stamens.   The five outer petals are
large and hooded, and in each is cradled a
bundle of four or more stamens.   These petals
and stamens are turned back; but there is a
second set of five petals which are generally
blotched with red, which stand erect, and en-
close a second set of stamens also erect, which
surround the style.

---

ORDER LXXXVIII.—TURNERIACEÆ.

The only genus in British gardens is Turnera,
and the species are hothouse and greenhouse
herbaceous plants, with flowers very like those
of the Bladder Ketmia.   On examination, however, it will be immediately seen that they do
not belong to the Mallow tribe, as their stamens
are distinct, whereas those of all the Malvaceæ
are united into a central column.

ORDER LXXXIX.—PORTULACEÆ.—THE PURSLANE
TRIBE.

The ornamental plants belonging to this
order, are all included in the genera Calandrinia, Portulaca, Talinum, and Claytonia ; and
those belonging to the first two of these genera
have very showy flowers.    In all the species the
flowers have a distinct calyx, generally of only
two sepals, which remains on till the seeds are
ripe ; and a corolla of five regular petals, which
close in the absence of the sun.    Each flower
has numerous stamens, and a single style with a
broad-lobed stigma which, is succeeded by a dry,
one-celled capsule, with a central placenta, to
which are attached numerous seeds.    The capsule opens naturally when ripe by splitting into
three or four valves.    But the most distinctive
mark by which plants belonging to this order
can be distinguished from others with simi-
larly shaped flowers, is their remarkably thick
fleshy leaves, an example of which may be seen
in the leaves of *Calandrinia discolor ;* and these
succulent leaves render all the ornamental
plants belonging to the order peculiarly liable to
be destroyed by frost or damp.    Some botanists
make a second order out of the plants usually
included in Portulaceæ, to which they give the
name of Fouquieraceæ.

## ORDER XC.—PARONYCHIEÆ.

WEEDY plants, containing among other genera, Knot-grass (*Illecebrum*), and Strapwort
(*Corrigiola*).	The new order Scleranthaceæ
has been separated from this; and it takes its
name from the British weed, Knawel (*Scleran-
thus*).

---

## ORDER XCI.—CRASSULACEÆ.—THE HOUSELEEK
TRIBE.

THE common Houseleek (*Sempervivum tecto-
rum*) grows, as is well known, on the tiles of
houses, or on walls, where there does not appear
a single particle of earth to nourish its roots.
The leaves are, however, so contrived as to
form a cluster of flat scaly circles, and thus to
shade and keep moist the roots beneath them.
The flowers, which are produced on a tall
flower-stem rising from the leaves, are pink,
and usually consist of a green calyx, cut into
twelve segments, and a corolla of twelve petals,
with twelve stamens and twelve carpels, which
spread out like a star in the middle of the
flower.	The number of petals, &c., is by no
means constant, as it varies from six to twenty;
but the other parts of the flower vary in the
same manner, and always agree with each other,

except as regards the stamens, which are sometimes twice the number of the petals, and arranged in two series, those in one series being abortive. At the base of each carpel is a kind of scale or gland, and this is the case with most of the genera included in the order. There are several species of Sempervivum, natives of the Canary Isles, which are very ornamental, and which have yellow flowers; but this genus, and that of Sedum, the Stone-crop, have been lately remodelled by Mr. Philip Barker Webb, and some new genera formed out of them. The principal other genera in the order are Crassula and Kalosanthes; the latter having been formed out of the former, and including those species of Crassula which have a tube-shaped corolla, with a spreading limb, divided into five segments, while the flowers of those species which have been left in Crassula have five distinct petals. All the plants belonging to the order have succulent leaves; and in all of them the number of the petals, sepals, and carpels, is the same, and of stamens either the same, or twice as many. In the common Houseleek, the anthers sometimes produce seeds instead of pollen.

CHAP. I.] FICOIDEÆ. 367

## ORDER XCII.—FICOIDEÆ.—THE FIG-MARIGOLD TRIBE.

THE principal genus in this order is that of Mesembryanthemum, the Fig-marigold. In the species of this genus, the leaves are always thick and fleshy, and sometimes in very singular shapes; and sometimes they are covered with a sort of blistery skin, which makes them look as though covered with ice, as in the Ice-plant (*M. crystallinum*). The leaves, when this is the case, are said to be papulose. Some of the species are annuals, others shrubby, and others perennials; and they are all natives of the Cape of Good Hope. The flowers, which are generally showy, have a green, fleshy, tubular calyx, with a four or five cleft limb, the tubular part of which encloses the ovary; and a corolla of numerous very narrow petals, which are arranged in two or more series. The stamens are very numerous; and the capsule has four or more cells, each of which contains numerous seeds. The valves of the capsule open when the seeds are ripe, if the weather should be dry; but remain firmly closed, so long as the weather continues wet.

The genera Reaumuria and Nitraria, which were formerly included in this order, have been removed from it, and made into separate orders,

the latter of which is introduced here; and the
genus Grielum, which was formerly included in
Rosaceæ was first removed to Ficoideæ, and
afterwards made into a separate order, under
the name of Neuradiaceæ, which precedes Ni-
trariaceæ.

---

### ORDER XCIII.—CACTACEÆ.—THE CACTUS TRIBE.

THERE is perhaps no order in the vegetable
kingdom which embraces plants so singular in
their forms as those comprehended in this
tribe. All the genera, with the exception of
Pereskia, are destitute of leaves, but they have
all succulent stems which answer the purposes
of leaves. The flowers of all the genera are
extremely showy; the calyx and corolla are
coloured alike, and confounded together; the
stamens are numerous, with versatile anthers
and very long filaments; the style is generally
long and slender, and the stigmas are numerous,
and either spreading or collected into a head.
The ovary is in the tube of the calyx, and it
becomes an eatable fruit, very similar to that of
the gooseberry. The genera are all natives of
tropical America. The principal kinds are the following: viz. Mammillaria, the stems of which
are subcylindrical, and covered with tubercles,

which are disposed in a spiral manner; and
each of which is crowned with a little tuft of
radiating spines mixed with down. The flowers
are without stalks, and they are disposed in a
kind of zone round the plant. The Melon
Thistle or Turk's-cap (*Melocactus communis*)
has a globose stem with deep furrows, the projecting ribs having tubercles bearing tufts of
spines. The stem is crowned with a woolly
head, from which the flowers are protruded,
the flowers themselves resembling those of
Mammillaria, but being larger. The Hedgehog
Thistles (*Echinocactus*) have stems resembling
those of the different species of Melocactus,
but they have not the woolly head; and the
flowers rise from the fascicles or tufts of spines
on the projecting ribs. The Torch-Thistle
(*Cereus*) has generally an angular stem with
a woody axis, and it has tufts of spines on the
projecting angles. It has not a woolly head,
and the flowers, which are very large and
showy, either arise from the tufts of spines, or
from indentations in the angles. The limits of
this genus are very uncertain; and several
plants which are included in it by some botanists,
are placed in other genera by others. The Old-
man Cactus was once called *Cereus senilis*, but
it is found to have a woolly head of great
size, which has very much the appearance of

a sable muff, and as, consequently, it cannot
belong to that genus, it has been called Pilo-
cereus. This plant is covered with long white
hairs, and, when of small size, it looks very
much like an old man's head. In its native
country, however, it grows to a great height,
and specimens have been imported fifteen feet
long, and not more than a foot in circumference.
The Peruvian Torch-Thistles (*C. hexagonus* and
*peruvianus*), in their native country, are upwards
of forty feet high, though not thicker than a
man's arm. They grow close together without
a single branch, and form a singular sort
of prickly crest on the summit of some of the
mountains in South America. The creeping
Cereus (*C. flagelliformis*) has slender cylindrical
trailing stems, which hang down on every side
when the plant is grown in a pot. The flowers,
which are very numerous, are pink. The night-
flowering Cereus (*C. grandiflorus*) only opens
during the night, and fades before morning;
the rays of the calyx are of a bright yellow when
open, and the petals are snow-white. The stem
is angular, branched, and climbing, throwing out
roots at every joint. The common Torch-thistle
(*C. speciosissimus*) is an erect plant, with a
three or four angled stem, and very large
bright crimson flowers, which are purplish
inside; and *C. speciosa*, sometimes called *Epi-*

*phyllum phylanthoides*, has thin leaf-like stems
with beautiful pale rose-coloured flowers. *C.
Jenkinsonii* is a hybrid between the last two
species. *C. truncatus* is another well-known
species. Opuntia has stems consisting of round,
flat, leaf-like bodies, united together by joints,
and generally covered with tufts of spines. The
most remarkable species are *O. communis*, the
Prickly Pear, grown to a great extent in the
South of Europe, and also in Brazil, as hedges,
the fruit of which is very good to eat; *O. Tuna*,
the Indian Fig, common in South America, and
much cultivated there, both as a hedge plant
and for its fruit; and *O. cochinillifera*, the
Nopal-tree, very much cultivated in Mexico
and South America, for the cochineal insect,
which feeds upon it. Rhipsalis has slender
cylindrical jointed stems, which look like sam-
phire. All these genera have only leaves when
quite young, and as soon as the plants begin to
grow, the leaves fall off. Pereskia, however,
is a genus belonging to this order which has
leaves like ordinary plants, which it retains
during the whole period of its existence. The
principal species are *P. aculeata*, the Barbadoes
Gooseberry, and *P. Bleo*, which has beautiful
rose-coloured flowers.

## ORDER XCIV.—GROSSULARIEÆ.—THE GOOSE-BERRY TRIBE.

This order consists of only one genus (*Ribes*), which includes all the Gooseberries and Currants; the two kinds forming two distinct sections. The first section, which embraces all the Gooseberries, has prickly stems, and the flowers are produced singly, or in clusters of not more than two or three together. The flower of the common Gooseberry (*Ribes Grossularia*) consists principally of the calyx (*a* in *fig.* 136), the five segments of the limb of which are turned back,

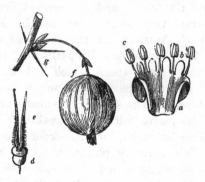

Fig. 136.—The Gooseberry.   (*Ribes Grossularia*).

and coloured of a reddish-brown. The petals (*b*) are white and erect, and bearded at the throat; but they are so small and inconspicuous, that

few people would notice them if they were not
pointed out. The stamens (*c*) are five in number, and erect, and the anthers burst lengthways on the inside. The ovary (*d*) is below the
cup of the calyx, and the style, which is cloven
to the base (*e*), is always covered with hairs in
the common Gooseberry (*R. Grossularia*), and is
more or less hairy in the other species. There
are two little bracteoles (*f*) on the pedicel ; and
a large bract, deeply cut, at the point from which
the pedicel springs (*g*). The leaves, which are
omitted in the engraving, also grow from the
same bud, and are three or five lobed, and
hairy ; and there are three spines just below
them. The fruit is a many-seeded berry, with
the seeds immersed in pulp ; and on cutting
open an unripe fruit, it will be found that the
seeds are each inclosed in an aril, with a separate footstalk, by which they are attached to a
membrane lining the sides of the berry, and
which is called a parietal placenta. The segments of the calyx remain on the ripe fruit.
Several of the ornamental species of Ribes belong to this division, as, for example, *R. triflo-*
*rum*, which has white flowers ; and *R. speciosum*,
which has crimson flowers, with the segments of
the calyx not reflexed, and long projecting
stamens like those of the Fuchsia. The fruit
and the whole of the stems and branches of this

species are covered with spines, and thus the
plant is easily distinguished from the common
gooseberry, the stem of which has no spines,
except three just below each bud.

The Currants are distinguished by the stems
being entirely without spines, and the flowers
being produced in racemes.   The leaves are cordate, and bluntly three or five lobed, a little
downy beneath, but smooth above. The flowers
of the Red Currant (*Ribes rubrum*) are numerous, and they are produced in drooping racemes,
with a little bracteole at the base of each footstalk (see *a* in *fig.* 137).   The calyx is flattish,
with the segments (*b*), which are of a pale
greenish colour, spreading widely, and not recurved.   The anthers (*c*) are loosely attached
to the filaments, and they burst sideways and

FIG. 137.—THE RED CURRANT.

across.   The style (*d*) is short, and divided into
two spreading stigmas at the apex.   The fruit
is smooth and transparent, with many seeds,
and it retains the remains of the calyx (*e*) when
ripe.   The white, and the striped or flesh-

coloured Currants, are varieties of *R. rubrum.*
The Black Currant (*R. nigrum*) has a more compact, and campanulate flower (see *a* in *fig.* 138),

FIG. 138.—THE BLACK CURRANT.

with the segments of the calyx reflexed ; the
anthers (*b*) are more firmly attached to the
filament ; the style (*c*) is not cleft, and the
stigma is two-lobed and capitate.  The fruit (*d*)
has a thick opaque skin, and the eye of the
calyx is larger ; the leaves are also covered on
the under surface with glands or cells, filled with
a fragrant oil formed by the limb, as shown at
(*e*), which represents the appearance of the leaf
when held up to the light.  There is often a solitary flower on a separate pedicel, at the foot of
the raceme ; and there are frequently ten stamens instead of five, and no petals, the petals having been changed into stamens—a metamorphose
the reverse of that which generally takes place.
　　The most ornamental kinds of Currant are
*R. multiflorum,* with very long drooping racemes

of greenish flowers ; *R. sanguineum*, the flowers
of which are crimson, and somewhat tubular;
*R. aureum*, which has the flowers of a golden
yellow, and quite tubular ; and *R. cereum*, which
has roundish leaves covered with white waxy
dots on their upper surface, and racemes with
few flowers, which are rather large, and of a
pure white. A few species, such as *R. saxatile*
and *R. Diacantha*, appear to be intermediate
between the Currant and the Gooseberry, as
they have the racemes of fruit common to the
one, with the spines and habit of growth of the
other. There is said to be another species
nearly allied to *R. sanguineum*, with dark-purple
flowers, which has not yet been introduced.

---

ORDER XCV.—ESCALLONIACEÆ.

Of the genera included in this order (which
were formerly included in Saxifragaceæ), Escal-
lonia is the most important, as it contains several species of ornamental South American shrubs.
The flowers of the different species vary considerably : in *E. rubra*, they are produced singly,
and the corolla, which is pink, is tubular, with
a short, five-cleft limb ; but in *E. montevidensis*
the flowers, which are white, are produced in
panicles, and have spread petals. The flowers

of both species have five stamens, and two carpels, the styles of which are combined. The
leaves are simple, alternate, and without stipules. Of the other plants contained in the
order, I may mention that *Itea virginica* is a
North American shrub, with white flowers;
and *Anopteris glandulosa*, which is also a shrub
with white flowers, is a native of Van Diemen's
Land.

---

### ORDER XCVI.—SAXIFRAGACEÆ.

THE genus Saxifraga of Linnæus has been
divided so as to form several genera; but they
do not appear to be generally adopted. The
flowers of all the species are rather small, and
they are generally racemose, or panicled; and
the corolla consists of five spreading petals with
short claws, and there are twice that number of
stamens. Among the most common species
may be mentioned London Pride (*Saxifraga* or
*Robertsonia umbrosa*), and the Meadow Saxifrage
(*Saxifraga* or *Leiogyne granulata*), the flowers of
the latter being large, and produced singly.
In the genus Hydrangea the flowers are disposed in corymbs, and they have five petals,
ten stamens, and from two to five styles; but
in the outer flowers of the corymb the stamens
and pistil are often wanting.

The genera Galax and Francoa, which were first included in Crassulaceæ, and afterwards in Saxifragaceæ, are now made into a new order called Galacineæ, or Francoaceæ, which is introduced here.

---

### ORDER XCVII.—CUNONIACEÆ.

This order, which was separated from Saxifragaceæ by Dr. Brown, contains principally hothouse plants with erect spicate racemes or panicles of small flowers. Weinmannia, Bauera, and Cunonia are the principal genera.

---

### ORDER XCVIII.—UMBELLIFERÆ—UMBELLIFEROUS PLANTS OR THE PARSLEY TRIBE.

This is a very large order, but it is so natural that no person who has seen Parsley in flower can ever be in any doubt as to an umbelliferous plant. Most of the species are either culinary plants, such as the Parsnep and Carrot, Celery, Parsley, Fennel, &c., or poisonous weeds, such as Hemlock, and the Water Parsnep; and there are very few ornamental plants included in the order: among these few may, however, be mentioned *Didiscus* or *Trachymena cærulea*, *Eryngium*, and *Bupleurum* or *Tenoria fruticosum*, *Angelica*, and *Heracleum*. Some of the species of the latter, particularly the Gigantic Siberian

Cow Parsnep (*H. asperum*), are perfectly magnificent objects. Notwithstanding the ease with
which these plants may generally be recognised,
as in some of the allied orders the flowers grow
in umbels or cymes, it may be necessary to remark
that Dr. Lindley defines umbelliferous plants
to consist of those which have their " flowers
growing in umbels, with inferior fruit, which,
when ripe, separates, or may be separated, into
two grains." Thus the common Dogwood is
not an umbelliferous plant, though its flowers
grow in umbels, because its fruit is a berry.

## ORDER XCIX.—ARALIACEÆ.

THE most interesting plant in this order is
*Hedera Helix*, the common Ivy; a well-known
climbing evergreen shrub, which throws out
roots from its branches at intervals, which it
strikes into any substance to which it can adhere. The flowers have all their parts in five
or ten divisions ; even the lower leaves, which
are smooth and leathery, are five-lobed. The
leaves on the flowering branches, which are
always in the upper part of the plant, are entire. The flowers are produced in umbels, and
they are succeeded by berries, which, in corre-
spondence with the parts of the flowers, are five
or ten celled. The large-leaved variety, called

the Irish Ivy, is a native of the Canary Isles;
and the gold and silver leaved, and golden
berried, are all varieties of the common kind.
There are, however, many exotic species, most
of which have not yet been introduced. The
genus Aralia, known by its two garden species,
*A. spinosa* and *A. japonica*, belongs to this
order. The first of these is called the Angelica
Tree, and is an old inhabitant of our gardens;
but *A. japonica* is of quite recent introduction.

---

### ORDER XCIX.*—HAMAMELIDEÆ.

THE most interesting plants in this order are
the Witch Hazel (*Hamamelis virginica*), and *Fothergilla alnifolia*.     In the first of these plants,
there are four long narrow petals, and the calyx
is four-lobed; and there are eight stamens, of
which four are fertile, and four barren. There
are two styles, and the capsules are leathery
and two-celled, and two-valved, with one seed
inclosed in an aril in each cell. The Witch
Hazel has the peculiarity of coming into flower
when it drops its leaves in autumn, remaining
in flower all winter, and forming its fruit in
spring, just as it is opening its new leaves. The
flowers are yellow, and very pretty from their
great abundance, and the light feathery effect
produced by the great length and narrowness

of the petals. The leaves are rough and feather-nerved, like those of the Hazel. Fothergilla is a pretty little shrub with terminal spikes of white flowers with yellow anthers. which are sweet-scented and appear before the leaves.

----

## ORDER C.—CAPRIFOLIACEÆ, OR THE HONEY-SUCKLE TRIBE.

This order, as originally constituted, may be divided into three tribes, viz., *Corneæ*, containing *Cornus*, *Benthamia*, and, according to some, *Aucuba ; Sambuceæ*, containing *Sambucus* and *Viburnum ;* and *Lonicereæ*, containing *Symphoria*, *Caprifolium*, *Lonicera*, *Leycesteria*, *Linnæa*, &c. Cornus, Benthamia, and some other genera, among which Dr. Lindley places Aucuba, are now formed into a separate order, under the name of *Cornaceæ.* The different species of Dogwood *(Cornus)* are known by the smooth bark of their stems and branches, which is frequently red, or reddish brown ; by their white flowers, which are produced either in heads, or umbels, or in corymbose panicles ; by their red or blackish berries ; and by their coarse feather-nerved leaves. The principal species of Cornus are the wild or female Cornel (*C. sanguinea*) ; the common Dogwood (*C. alba*) ; the male Cornel, or Cornelian Cherry (*C. mas*) ; and American

Dogwood (*C. florida*). All these plants have a very small four-toothed calyx, and a corolla of four petals. There are four stamens and one style. The fruit is a berry-like drupe. Some of the species, as for example *C. florida*, have a large involucre of four leaves, having the appearance of petals. *Benthamia fragifera*, called by Dr. Wallich *C. capitata*, has an involucre of four leaves of yellow, tinged with red, surrounding a head of small greenish inconspicuous flowers. The fruit consists of a number of drupes, grown together like a Mulberry, with six, eight, or more seeds, surrounded with a viscid pulp. The leaves are long and tapering, of a fine texture, and of a light green above, and silvery white below.

The genus *Sambucus*, the Elder, is characterised by its pinnate leaves and terminal cymes of flowers, which have a small five-lobed calyx, a rotate corolla also five-lobed, five stamens about the length of the corolla, no style, and three obtuse stigmas. The berries are globular, pulpy, and one-celled; each containing three or five seeds, which are convex on the outside, and angular within. The berries differ in colour in the different species, those of the common kind being a deep purplish black, and those of *S. racemosa* being red. The stems and branches are of a soft wood, having a white spongy pith. The white-berried Elder is a variety of the common kind.

The genus *Viburnum* contains several well-known plants, among which may be mentioned the Laurestinus (*V. Tinus*), the Guelder Rose (*V. Opulus*), and the Wayfaring Tree (*V. Lantana*). This genus is very nearly allied to Sambucus in the flowers, but it is easily distinguished, on examination, by its leaves, which are not pinnate, and by its wood, which is hard and not spongy. The berries have also only one seed, and they are not eatable,— those of the Laurestinus are, indeed, injurious. The Laurestinus and some other species are evergreen; but by far the greater number of species are deciduous.

The genus *Lonicera* formerly included all the kinds of Honeysuckle; but now only the upright species, or what are called the Fly Honeysuckles, are comprised in it, and the climbing kinds are called Caprifolium. One of the upright kinds, most common in gardens, is the Tartarian Honeysuckle (*L. tartarica*), the flowers of which are in twins. The corolla is tubular and funnel-shaped, with a five-cleft limb. There are five stamens, a filiform style, and a capitate stigma. The berries are distinct when young, but they afterwards grow together at the base. The leaves are always distinct. The genus *Caprifolium* embraces all the climbing species, the flowers of which are disposed in whorls, and the upper leaves are connate, that is, growing together at the base, so that two appear only one leaf, with the stem passing through it. A single leaf of this kind is called perfoliate. The flowers spring from the axils of the leaves, and are what are called ringent, that is, they are composed of five petals, four of which grow together, almost to the tip, while the fifth is only attached to the others about half its length, and has the loose part hanging down. Flowers of this kind, with their lower part forming a tube, and their upper part widely open, are said to be gaping. In the Trumpet Honeysuckle (*C. sempervirens*) the tube of the corolla is very long, and the lobes of the limb nearly equal; and the flowers, instead of springing from the axils of the leaves, form terminal spikes, each consisting of three or more whorls of flowers.

The Snowberry (*Symphoria racemosa*) bears considerable resemblance to the upright Honeysuckles. The flowers are funnel-shaped, and four or five lobed. The berry has four cells, but two of the cells are empty, and the others have only one seed in each. The leaves are oval, quite entire, and not connate.

*Leycesteria* is a very handsome shrub, with

white flowers, and very large and showy purple and reddish bracts. The berries are of a very dark purple, and they are nearly as large as a gooseberry. *L. formosa* is a native of Nepaul, but it appears tolerably hardy in British gardens, and it stands the sea-breeze without injury.

*Linnæa borealis* is a little innsigificant trailing plant, which is included in this order, and which is only worth mentioning on account of its being named in honour of Linnæus. It is a half-shrubby evergreen, with small bell-shaped flesh-coloured flowers, which are said to be fragrant at night.

## ORDER CI.—LORANTHEÆ.

Four genera are included in this order, all remarkable in different ways. The first of these is the common Mistletoe (*Viscum album*), a most remarkable parasite, a native of Britain, and generally found on old apple-trees ; and the second is *Loranthus europæus*, a native of Germany, closely resembling the Mistletoe, but found generally on the oak, where the true Mistletoe rarely grows. This plant is said to have been introduced in 1824, but it is not now in the country. There are other species of the genera, one a native of New Holland. *Nuytsia floribunda*, also a native of New Holland, a very curious plant, is also included in this order. It is a shrub about three feet high, so covered with orange-coloured blossoms that the colonists call it the Fire-tree. When the seed of this plant germinates, it is said to have three cotyledons. The last plant generally included in this order is *Aucuba japonica*, though it is probable this plant belongs to Cornaceæ. Of this species we have probably only a variety, from the variegation of the leaves; and it has never produced seeds, as only the female plant has been introduced.

### ORDER CII.—CHLORANTHEÆ.

INCONSPICUOUS plants with greenish flowers, which require a hothouse in Britain.

### ORDER CIII.—RUBIACEÆ. (SEE CHAP. V. P. 85.)

THIS order is divided into thirteen sections, most of which have been already described. In all the species the tube of the calyx adheres to the ovary, which is crowned with a fleshy cup, from which arises the single style; and the petals are united at the base, and attached to the upper part of the tube of the calyx.

### ORDER CIV.—OPERCULARIEÆ.

EXOTIC weeds, formerly included in Rubiaceæ.

---

### ORDER CV.—VALERIANEÆ.— THE VALERIAN TRIBE.

No person can ever have been in the neighbourhood of Greenhithe, in Kent, without having observed the red Valerian, which grows in such abundance on the steep banks of the chalk-pits in that neighbourhood; and probably still more of my readers will be familiar with the common wild Valerian, or All-heal, which is found in moist places, generally among sedges, in every part of England. Another species of the same genus is common in Scotland, so that the name of Valerian is familiar to all persons who know anything of British plants. Common as these plants are, however, probably most of my readers are unaware of the very curious construction of their flowers; or of the very great variety exhibited by the different species. The genus Valeriana is, indeed, one which presents a remarkable instance of variety of construction, united with a similarity of form which makes all the species recognisable at a single glance. In all the species, the corolla is funnel-shaped, with a

long tube, and a five-lobed limb. In the red
Valerian (*V. rubra*), the lower part of the tube
is drawn out into a spur; and on this account
the plant is sometimes called the spurred Valerian, and it has been placed by De Candolle
in a new genus, which he called Centranthus.
The other species of Valerian have the tube of
the flower gibbous, that is, much larger on one
side than on the other. In all the calyx is
tubular, with the limb curiously rolled, so as to
form a rim or crown to the fruit, like that on
the heads of basket-women. When the flowers
drop, the fruit, which is one-celled and one-
seeded, and which adheres closely to the tube
of the calyx, begins to swell, and as it does so
the limb of the calyx gradually unrolls, till at
last, when the fruit is ripe, it forms a sort of
feathery tuft to waft it away. The leaves of
plants of this genus vary exceedingly, even on
the same plant; but generally those of the red
Valerian are lanceolate; those of *V. dioica* are
pinnatifid; those of the wild Valerian (*V. officinalis*), pinnate; and those of the garden Valerian, the kind found in Scotland, (*V. pyre-
naica,*) are cordate. The flowers of *V. dioica*
are male and female, and are found on different
plants. The principal other genera in this
order are Valerianella, the Corn Salad or
Lamb's Lettuce; and Fedia, the Horn of
Plenty.

## ORDER CVI.—DIPSACEÆ.—THE TEASEL TRIBE.

THE principal genera belonging to this order
are Dipsacus, the Teasel, and Scabiosa, the Scabious; to which may be added a pretty little
annual called Knautia. The plants belonging
to this order bear considerable resemblance to
those included in Compositæ, as they consist of
a head of florets seated on a common receptacle,
which is chaffy, and surrounded by an involucre.
The florets are also furnished with what may be
called a double calyx, the limb of the inner
part being cut into long teeth, and resembling
the pappus of the Compositæ. In the genus
Dipsacus, the most important plant is the Ful-
ler's Teasel (*Dipsacus fullonum*), in which the
receptacle is raised in the form of a cone, and
the chaffy scales are hooked, and so strong,
that the flower-heads when dry are used for
preparing broad-cloth. The leaves of this plant
are opposite, and united at the base. The
florets have a four-cleft corolla, and four distinct stamens; differing in this respect decidedly from the Compositæ, which have five
stamens, the anthers of which are always united
into a tube. *Dipsacus sylvestris* might be easily
mistaken for a kind of Thistle; but the difference will be seen at once by examining the
anthers of the florets. The Devil's-bit Scabious,

which is so called from the root looking as though
a part had been bitten off, has the same kind
of flower-head as the Dipsacus, but the receptacle is flat, and the involucre much smaller.
In some of the species of Scabious, the florets
of the outer ring resemble those of the ray in
flowers of the Compositæ. The leaves of the
genus Scabious are as variable as those of the
genus Valeriana, scarcely two species being
alike.

-----

### ORDER CVII.—CALYCEREÆ.

Obscure American plants, nearly allied to
Compositæ.

-----

### ORDER CVIII.—COMPOSITÆ. (See Chap. VI. P. 98.)

Plants with heads of florets on a common
receptacle, surrounded by an involucre. The
florets are of three kinds, viz., ligulate, tubular,
and bilabiate ; the heads consisting sometimes
entirely of florets of one kind, and sometimes
with ligulate florets forming the outer ring,
called the ray, and tubular flowers forming the
centre, called the disk. The calyx continues on
the ripe fruit, and its limb is frequently cut into
a kind of fringe called the pappus. The fruit
is of the kind called an achenium, that is, dry
and bony, and continuing enveloped in the persistent calyx, but without adhering to it.

ORDER CIX.—LOBELIACEÆ.

THE genus Lobelia is well known from the pretty little blue-and-white flowering plants that are so common in pots for windows and balconies, and that continue flowering so freely all the summer. There are two or three species which are grown for this purpose, viz. *Lobelia Erinus*, *L. bicolor*, and *L. gracilis*, all annuals, which require to be raised on a hotbed by sowing in February, and which will then flower all the summer, with no other care than regular watering. All these flowers have the tube of the calyx united to the ovary, with a five-parted limb. The corolla is irregular and tubular, with the tube cleft on the upper side, and thickened at the base. The limb of the corolla is divided into two parts ; one of which, called the upper lip, is cut into two narrow sharp-pointed segments, which stand erect; while the lower lip, which is much the longer, and hangs down, is cut into three rounded segments. There are five stamens, the anthers of which grow together, and at least two of them are bearded. The capsule is oval, two-celled, two-valved, and many-seeded, opening naturally at the top when ripe. These general characters will be found in all the numerous species of Lobelia, as the genus at present stands, as they

all have the two hornlike segments of the upper
lip, and the rounded lobes in the pendulous
under lip; and many of the plants formerly
called Lobelia which differ in these particulars
have been placed in other genera. Thus Tupa,
which contains several of the large scarlet-
flowered species, has the segments of the limb
of the corolla united at the tip; the filaments
of the stamens cohering as well as the anthers,
and the stigma protruding. Siphocampylos has
the tube of the corolla ventricose in the middle,
the segments of the upper lip long and curving
over each other, and the lower lip very slightly
lobed, with both the filaments and the anthers
combined. In Dortmannia the filaments are
free, and only the anthers combined; in Para-
stronthus (*L. unidentata*), there is scarcely any
tube to the corolla, and in Isotoma, the corolla
is salver-shaped. The beautiful little *Clintonia
pulchella* belongs to this order, and it differs
from Lobelia in its corolla having scarcely any
tube, and also, but more decidedly, in the very
long tube of its calyx. This is so long and slender as to look like a part of the flower-stalk; as
does the capsule, which, when ripe, is triangular,
and is as long as the silique of a cabbage or
wall-flower, to which it bears considerable resemblance. All the Lobeliaceæ have an acrid
milky juice, which is poisonous.

### ORDER CX.—STYLIDEÆ.

THIS order contains three genera of New Holland plants, only one of which has been introduced. The flowers are tubular, with a five-cleft limb, and they are covered with hairs, terminating in capitate glands; the stamens are united into a column, which is bent towards the fifth or lower segment of the limb, which is much larger than the others. The united stamens are so irritable as to start forward when touched with a pin.

### ORDER CXI.—GOODENOVIÆ.

ALL the plants in this order are natives of New Holland, and they bear considerable resemblance to those included in Lobeliaceæ, but they have not a milky juice, and the stigma, which is very small, and without any style, is surrounded by a curious cup called an indusium, which is generally found full of pollen. This very remarkable organ is probably rendered necessary by the very small size of the stigma, which can only absorb the pollen very slowly. The most interesting genera contained in this order are Lechenaultia and Euthales.

## ORDER CXII.—CAMPANULACEÆ.—THE
## CAMPANULA TRIBE.

THE plants in this order have a bell-shaped regular corolla, consisting of five petals, usually grown together so as to form a monopetalous corolla with five lobes, each lobe having a conspicuous central nerve or vein. There are five or more stamens, which are generally distinct, and which have broad bearded filaments bending over the ovary. The style is at first short, but it gradually elongates itself, and both it and the stigmas are furnished with tufts of stiff hairs, which, as the style pushes itself through the stamens, brush off the pollen, and retain it till the stigma is in a proper state to receive it. The anthers burst as soon as the corolla opens. The capsules have generally two, three, or five cells, and each cell contains many seeds.

In the genus Campanula, the capsule opens by little valves, which look as though cut with scissors. The juice of the plants is milky, but not poisonous. The principal genera are Campanula, Prismatocarpus (Venus's Looking-glass), Roellia, Phyteuma (the petals of which are distinct), Trachelium, Wahlenbergia, and Adenophora. Lobeliaceæ and Goodenoviaceæ were formerly included in this order.

### ORDER CXIII.—GESNERIEÆ.

THE corolla is tubular and sub-bilabiate, with a five-cleft limb. There are four stamens, two longer than the others, with the rudiments of a fifth. The anthers generally adhere in pairs; the fruit is one-celled and many-seeded; the leaves are thick and covered with a soft down; and the roots are frequently tuberous. The qualities are excellent. The species of the genus Gesneria are usually hothouse plants, with bright scarlet flowers; and those of Gloxinia have generally purple flowers; and of Sinningia the flowers are greenish.

### ORDER CXIV.—VACCINIEÆ. (SEE CHAP. VII. P. 130.)

THIS order includes the Whortle-berries, Bilberries, and Cranberries, and it is very nearly allied to Ericaceæ, from which it is distinguished by the disk, which lines the calyx, entirely surrounding the ovary, which is thus placed below the rest of the flower, and is called inferior. The fruit is a berry.

### ORDER CXV.—ERICACEÆ. (SEE CHAP. VII. P. 109.)

ALL the Heath tribe, including the Arbutus, Rhododendron, Azaleas, &c., are distinguished by their anthers, which have a little hole or pore at the apex of each cell; each cell being also generally furnished with a kind of spur at its base. The stamens in all these genera grow from beneath the ovary, and the filaments are thick and fleshy. The fruit is a dry capsule, or follicle.

---

### ORDER CXVI.—PENEACEÆ.

BEAUTIFUL shrubs, natives of the Cape of Good Hope, with the habit of Pimelea, and corymbs of pale pink flowers. The calyx is in two sepals, the stigma four lobed, and the fruit four-valved, with two seeds in each cell.

### § III.—COROLLIFLORÆ.

The plants comprised in this division are called monopetalous, as they have their petals joined together, so as to form a cup for the stamens and pistils quite distinct from the calyx; and the stamens are attached to the corolla.

---

### ORDER CXVII.—EPACRIDEÆ.

THIS order stands on debateable ground, being by many botanists included in the last division; but it seems properly placed in this,

as the stamens are attached to the petals,
which adhere together; and if a flower of any
species of Epacris be examined, it will be found
that the corolla, with the stamens attached to
the lining of the tube, parts readily from the
calyx without losing its natural form. The
flowers are tubular or campanulate, with a five-
cleft limb, and will divide readily into five
petals, each of which has the filament of a
stamen attached to it, leaving only the anthers
free. The anthers are one-celled and awnless,
and this is the principal distinction between this
order and Ericaceæ. The calyx is five-cleft,
coloured like the corolla; and there are five
scale-like bracts below it, which look like a
calyx. The capsule is dry, with the seeds attached to a central column. The leaves are dry,
hard, and prickly. The species are natives of
Australia, where they supply the place which
the Heaths hold in Europe and Africa; no
Heath having been yet found in any part of
Australia.

---

### ORDER CXVIII.—SYMPLOCINEÆ.

THIS order contains one genus, Symplocos, of
greenhouse and stove shrubs, from South America, with small white flowers, and serrated
leaves, which turn yellow in drying.

## ORDER CXIX.—STYRACINEÆ.

THE plants in this order best known in
English gardens are *Styrax officinale*, the Storax,
and *Halesia tetraptera*, the Snowdrop-tree. The
flowers of both are white ; those of Storax are
funnel-shaped, with a five-cleft limb ; there
are ten stamens, growing together at the base,
with short filaments, and very long anthers.
The fruit is a drupe which is nearly dry, containing a one-celled nut, enclosing from one
to three seeds. The seeds have two skins, the
inner one like a cobweb, and the outer one
spongy. The bark, when wounded, affords the
gum called storax. Halesia has drooping bell-
shaped white flowers, something like those of the
Snowdrop, (see *a* in
*fig.* 139,) with four
petals and twelve or
sixteen stamens com-
bined into a tube at
the base. The fruit is
a dry, winged drupe,
which has four an-
gles in *H. tetraptera*
(*h*), and two in *H.
diptera ;* and which
contains a stone or
putamen (*c*), which

FIG. 139.—SNOWDROP TREE (*Halesia
tetraptera*).

has two or four cells, and as many seeds. Some
botanists make Halesiaceæ a separate order.

---

### ORDER CXX.—MYRSINEÆ.

SHOWY shrubs, with evergreen leaves, and
cymes of white or red flowers, which require a
stove or greenhouse in England. The plants
belonging to this order may be easily known on
cutting open their flowers, as they are the only
monopetalous flowers among the stove plants
that have the stamens opposite the lobes of the
corolla ; the general position of the stamens
being between the lobes. The principal genera
in this order are Myrsine, the species of which
are greenhouse shrubs ; and Ardisia, the latter
being well-known stove shrubs, with white
flowers and red berries. Theophrasta, Clavija,
and Jacquinia, were included in this order; but
they are now formed into a new one, under the
name of Theophrasteæ.

---

### ORDER CXXI.—SAPOTEÆ.

THIS order is best known by the genera *Arga-
ria*, *Sideroxylon*, *Chrysophyllum*, and *Bumelia*,
all of which are stove or greenhouse plants.
The seeds of *Achras Sapota* contain abundance of

oil, which is so concrete as to have the appearance of butter ; and hence the tree is called the
Butter-tree.  *Sideroxylon* has such hard wood
as to be called the Iron-tree.   The juice of all
these plants is milky, and the milk is wholesome
as food.

---

### ORDER CXXII.—EBENACEÆ.

THE principal genus is Diospyros ; which contains the Ebony-tree (*D. Ebenum*), the Date-
plum or Lotus-tree (*D. Lotos*), both natives of
the East Indies ; and the Persimon (*D. virginiana*), a native of North America.   The species
are trees with hard dark wood ; that of Ebony
is quite black when old, and remarkably heavy.
The flowers are white and inconspicuous, and the
fruit, which is eatable, but insipid, is a berry,
placed in the centre of the calyx, which spreads
round it like a saucer.   It is very harsh when
first gathered, and must be kept till it is half
decayed, like the Medlar, before it is eaten.

---

### ORDER CXXIII.—BREXIEÆ.

LARGE stove trees, with axillary branches of
white flowers.

ORDER CXXIV.—OLEINÆ.—(See Chap. VIII. P. 136.)

THIS order comprises the common Ash, the Manna Ash, the Olive, the Privet, the Fringe-tree, the Phillyrea, and the Lilac. The flowers of all have only two stamens, and a roundish two-celled ovary, without any disk. The flowers of the Ash have no corolla, and the fruit is a samara. In the other genera, the flowers are more or less funnel-shaped, and the fruit is a capsule. The leaves are generally pinnate, and always opposite. The seeds have a dense albumen.

ORDER CXXV.—JASMINEÆ.—(See Chap. VIII., P.134.)

THIS order has been separated from the last, chiefly on account of the seeds having no albumen. The principal genus is the Jasmine, which has a funnel-shaped corolla, and pinnate leaves. Some botanists insert, between Oleaceæ and Jasmineæ, the new order Columellieæ, which contains only one plant, *Bolivaria trifida*.

### ORDER CXXVI.—STRYCHNEÆ.

TROPICAL trees. The principal genus *Strychnos*, the fruit of which is the well-known poisonous nut, *Nux vomica*. The genera Theophrasta and

Fagræa were formerly included in this order;
but the first is now placed in the new order,
Theophrasteæ (see p. 399); and the second is
placed in another new order introduced here,
and called Potaliaceæ.

---

### ORDER CXXVII.—APOCYNEÆ.

VERY showy plants from various parts of the
world, some of which require a stove in Britain,
while others are quite hardy. They also vary
in some of them being trees, others erect shrubs
or climbers, and others perennial; but they
are all easily recognised by the twisted direction
of the segments of the corolla, which has been
compared to the rays of St. Catharine's-wheel.
The corolla is generally salver-shaped as in the
periwinkle (*Vinca major*), or funnel-shaped, as
in *Taberna montana*, and *Allamanda cathartica*,
or divided into equal segments as in *Nerium
Oleander*. The flowers are often bearded in the
throat, and furnished with hypogynous scales;
with the stamens inclosed in the flower, and the
anthers lying close together. The seed is contained in two follicles, which are slender, and have
their seeds disposed in two rows. The species all
abound in an injurious milky juice; and two of
the genera, Cerbera and Tanghina, are virulent
poisons.

ORDER CXXVIII.—ASCLEPIADEÆ.

THESE plants are very nearly allied to the last, and they differ chiefly in having the segments of their corollas straight, in their stamens being united into a sort of crown, and in their pollen being found in masses of a waxy substance. The seeds are also each furnished with a tuft of fine long silky hair. The principal plants are *Periploca græca*, a hardy, climbing, shrub, with rich, dark, velvet-looking flowers, which are said to be poisonous to flies, and *Hoya carnosa*, a stove or greenhouse climber, with waxen-looking, clustered, odoriferous flowers, distilling honey; to these may be added Pergularia, a stove climber, remarkable for its fragrance, Physianthus, Gonolobus, Ceropegia, and Asclepias, all singular-looking climbing plants; and besides these, I may mention Stapelia, the species of which are dwarf plants, with their flowers hanging down below the pots in which they grow, and the odour of which is so like that of carrion, as to induce flesh-flies to lay their eggs upon them.

ORDER CXXIX.—GENTIANEÆ.—THE GENTIAN-TRIBE.

THE best known genera are Gentiana, (the Gentian), Lisianthus, and Menyanthes (the

Buckbean). The flowers have a tubular calyx
and corolla, the latter plaited in the tube, and
with an equally-parted limb, which is generally
five cleft; and an equal number of stamens
with broad filaments, and arrow-shaped anthers.
The seeds are numerous, and are usually in two
follicles.

The orders Spigeliaceæ, Loganiaceæ, and
Menyanthaceæ, have been separated from Gentianeæ, and are adopted by some botanists.

---

### ORDER CXXX.—BIGNONIACEÆ.

THE most interesting genera are—Bignonia;
from which Tecoma has been divided by some
botanists, on account of a slight difference in
the seed-pod; Jacaranda, said to produce the
rosewood of commerce; Eccremocarpus, and
Catalpa. All the plants included in this order
have winged seeds, and generally very long
horn-like seed-pods. The different species of
Bignonia or Tecoma have trumpet-shaped flowers
with a five-toothed calyx, and four stamens of
unequal length, with the rudiments of a fifth.
The capsule is very long and narrow, resembling
a silique in shape, but broad on the outside, and
the leaves are pinnate. *Eccremocarpus*, or *Calam-
pelis scabra*, is a well-known climber, with orange-
coloured, bag-like flowers, which are produced
in secund racemes; large, roundish warted

fruit, with winged seeds; and pinnate leaves, with tendrils. In Catalpa the corolla has a very short tube, and an unequal, five-lobed limb. There are five stamens (only two of which are fertile); and an exceeding long, cylindrical, silique-shaped seed-pod, which is sometimes two feet or more in length. The leaves of the Catalpa are heart-shaped. In Jacaranda, the capsule is above two feet long, and quite flat. *Crescentia cujète,* the calabash-tree, belongs to this order.

----

## ORDER CXXXI.—COBÆACEÆ.

THIS order is restricted to one genus Cobæa, of which one species (*C. scandens*) is common in British gardens. This plant is an annual climber, with showy bell-shaped flowers, which are first green, and afterwards become purple. This plant has remarkably long tendrils, which twist themselves round any thing that comes in their way.

----

## ORDERS CXXXII. AND CXXXIII.—PEDALINEÆ AND SESAMEÆ.

THESE orders are now united into one, under the name of Pedalineæ; and the most interesting genus is Martynia, consisting of half hardy annual plants with bell-shaped flowers, and very curious seed-pods.

ORDER CXXXIV.—POLEMONIACEÆ.

This is a very interesting order to the lovers
of ornamental flowers, from the beauty of those
of some of the genera.  The genus Polemonium,
the Greek Valerian, has one species (*P. cœruleum*)
which is found wild in many parts of England,
and is known by the names of Charity and
Jacob's Ladder.  The corolla, which is of a
deep blue, softening into white in the centre,
is rotate, with the stamens, which are bearded
at the base, inserted in the throat.  The capsule
is three-celled, and many-seeded, as is generally
the case with plants in this order, and the leaves
are pinnate.  The Phloxes are well-known; all
the species are very handsome, but none are
more so than the beautiful annual (*P. Drum-
mondi*).  The corolla of these plants is salver-
shaped, with an elongated tube, the limb twisted
in the bud, and wedge-shaped segments.  The
stamens are inserted above the middle of the
tube, and the cells of the capsule are one-
seeded.  Leptosiphon has the corolla funnel-
shaped, with a very long slender tube, and a
campanulate limb with oval lobes; the corolla
is covered with a great number of fine glandular
hairs, and the limb is twisted in the bud.  The
stamens, which have very short filaments, are
inserted in the throat of the corolla.  The calyx

consists of five sharply-pointed hairy lobes, connected by a very fine membrane. The flowers
are surrounded by a great number of sharply-
pointed bracts. Similar bracts are very conspicuous in the genus Collomia. Gilia and
Ipomopsis, so well known for their splendid
flowers, also belong to this order.

## ORDER CXXXV.—HYDROLEACEÆ.

ELEGANT little plants, distinguished from the
preceding order by the flowers having two
styles, and a two-valved capsule. Retziaceæ,
an order containing only one Cape plant, is
inserted here by some botanists, who have
separated it from Convolvulaceæ.

## ORDER CXXXVI.—CONVOLVULACEÆ.

THE principal genera are Convolvulus, Ipomœa
and their allies. The genus Convolvulus formerly
included all the beautiful monopetalous flowers
with a folded limb, which are so common in gardens, but it is now restricted to those which
have a two-celled capsule, with the cells two-
seeded ; the stamens are inclosed in the corolla,
and the stigma is divided into two narrow
thread-like lobes. Ipomœa only differs in having
the lobes of the stigma capitate. In Quamoclit,

the little scarlet Ipomœa, the capsule is four-
celled, and the cells one-seeded; the corolla is
tubular, and the stamens project beyond the
throat.     Batatas, the Sweet-potato, resembles
Quamoclit, but the corolla is campanulate, and
the stamens are inclosed.     In Pharbitis (in
which genus the common Convolvulus major,
and the beautiful Ipomœa Learii, are both now
included), the capsule is three-celled, and the
cells are three-seeded; and in Calystigia, in
which is now placed the common bindweed of
the hedges, the capsule is one-celled and four-
seeded; and the flower, which in other respects
agrees with that of the genus Convolvulus, has
two bracts which serve as a sort of involucre.
All these flowers have the lobes of the corolla
marked with a decided fold or plait, and they
are climbing plants, generally annuals. Cuscuta
is a parasite belonging to Convolvulaceæ, which
though it springs from the ground, withers
just above the root as soon as it has twined
itself round any plant within its reach; drawing its entire nourishment from the unfortunate plant it has attacked, and which it soon
kills.     The plants in this order produce an
acrid milk; and the roots of a kind of Convolvulus yield the drug called Jalap, which
takes that name from the Mexican city Xalapa,
near which it is grown.

## ORDER CXXXVII.—BORAGINEÆ.

THE fruit of the plants included in this order consists of four distinct carpels, each containing a bony nut. These nuts frequently appear as though a hole had been bored in them at the base, and they are frequently striped or twisted. The flowers are generally secund, or rather they are produced in spikes which appear to have flowers only on one side, from the spikes being curiously rolled up before the flowers expand, and uncoiling gradually as they open. The corolla is generally salver or funnel shaped, with a five-lobed limb, and five little scales just within the throat, which appear to be placed there to close up the orifice. There are five anthers, which seem attached to the corolla, without any stamens, and a slender style terminating in a two-lobed stigma. The calyx is tubular, and remains on the fruit till ripe; the teeth of the calyx contracting at the point, so as to cover the ripe carpels. The principal genera are Pulmonaria (Lungwort), Symphytum (Comfrey), Cerinthe (Honeywort), Lithospermum (Gromwell), Echium (Viper's Bugloss), Anchusa (Bugloss); Myosotis (Scorpion-grass or Mouse-ear), one species of which, *M. palustris*, is the Forget-me-not; Omphalodes (Venus' Navelwort), Cynoglossum (Hound's-tongue), and Heliotropium (the Heliotrope).

### ORDER CXXXVIII.—CORDIACEÆ.

EAST INDIA trees and shrubs of which Ehretia is, perhaps, the best known. Nearly allied to Boragineæ.

---

### ORDER CXXXIX.—HYDROPHYLLEÆ.

THIS order is interesting from its containing Phacelia, Eutoca, and Nemophila, all well known Californian annuals.

---

### ORDER CXL.—SOLANACEÆ (SEE CHAP. IX. P. 141).

THE genera Verbascum and Celsia have been removed from this order, and formed by some botanists into another called Verbascinæ, though by Dr. Lindley they are included in Scrophularinæ. The plants left in the order Solanaceæ have all a tubular calyx, which remains on the fruit till it is ripe; and the fruit itself is generally round and fleshy, with two or four cells and numerous seeds. In some of the genera, the permanent calyx looks like a capsule, but on opening it, the little berry-like fruit will be found inside. There are five stamens, the anthers of which are two-celled like those of most other plants, and the filaments are inserted in the corolla, which is generally partly tubular

with a spreading limb, the segments of which
are plaited, that is, each bears the crease of a
fold in the middle, as may be seen in the Petunia. In the order Verbascinæ, the corolla is
rotate, and the segments are not plaited ; the
anthers also are only one-celled. Most of the
plants belonging to Solanaceæ are poisonous in
a raw state ; but they lose their deleterious
qualities when cooked.

---

### ORDER CXLI.—SCROPHULARINÆ.

The Foxglove is generally taken as the type
of this order, and it has a tubular corolla (see *a*
in fig. 140) with a short limb (*b*), and a spreading calyx (*c*). There are four stamens of unequal length inserted on the base of the corolla

and hidden in its tube; and
an oblong ovary (*d*), with
a long style, and a two-
lobed stigma (*e*). The
fruit is a dry capsule with
two cells, and numerous
seeds. The flowers of the
other genera are very irregular. In the Snapdra-
gon, the corolla is what is
called personate ; and in

Fig. 140.—Foxglove
(*Digitalis*).

the Calceolaria the lower lip is curiously inflated.

The stamens also differ. In most of the genera there are four, but in Pentstemon there is a fifth, long and slender, and hairy at the point, but without any anther; and in Calceolaria and Veronica there are only two. Among the genera included in this order may be mentioned Buddlea, the flowers of which grow in ball-like heads; Paulownia, Maurandya, Mimulus, Alonsoa, and Collinsia. The Toadflax (*Linaria*), and several other British plants belong to it; but the Yellow Rattle (*Rhinanthus*), and some other allied plants, have been formed into a new order called Rhinanthaceæ; Chelone and Pentstemon have been formed into an order called Chelonaceæ; and Sibthorpia, Disandra, &c., into one called Sibthorpiaceæ. Trevirana or Achimenes, and Columnea, are removed to Gesneriaceæ.

The new order Cyrtandraceæ, including Æschynanthus, Streptocarpus or Didymocarpus, Fieldia, and Amphicoma, is introduced here: the first and last of these genera are new, and the others were formerly included in Bignoniaceæ.

---

### ORDER CXLII.—LABIATÆ.

THE plants belonging to this order include Mint, Sage, Thyme, and other kitchen aromatic plants, and several well-known British weeds.

They are all distinguished by a tubular, bilabiate corolla with a projecting under lip (see *a*
in fig. 141). In some plants the corolla is ringent, as shown in fig. 142, taken from Dr. Lindley's *Ladies' Botany*, in which *a* is the galea or

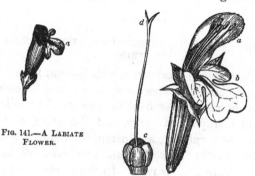

FIG. 141.—A LABIATE
FLOWER.

FIG 142.—BLACK HOREHOUND
(*Ballota nigra*).

helmet, and *b* the lower lip, which is three-
lobed. There are four stamens, two of which
are longer than the others, and the cells of the
anthers differ from those of most other plants
in spreading widely apart from each other, each
being joined to the filament only at the tip.
The pistil consists of four distinct carpels (*c*),
a very long style lobed at the tip, and furnished
with a very small stigma at the tip of each lobe
(*d*). The flowers of some of the plants belonging to this order are disposed in a whorl round

the stem; as, for example, those of the Dead
Nettle (*Lamium*).   Among the other plants
belonging to the order may be mentioned the
Bugle (*Ajuga*), and the Ground Ivy (*Glechoma*),
both common but very pretty British weeds.

---

### ORDER CXLIII.—VERBENACEÆ.—THE VERVAIN TRIBE.

THE genus Verbena is well known, from the
many beautiful species now common in every
greenhouse.   The fruit is two or four celled,
and a drupe or a berry, and the calyx of the
flowers is tubular, and persistent round it ; but
the corolla is deciduous, and falls off long
before the fruit is ripe.   In the genus Verbena
the calyx is tubular, with five distinct angles,
ending in five teeth.   The corolla has a cylindrical tube nearly double the length of the
calyx, and a flat limb divided into five unequal
segments, which are wedge-shaped and notched,
the central one of the lower three appearing to
have been slightly pinched ; the throat of the
corolla is hairy.   There are four stamens, two
longer than the others, the anthers having two
widely-spreading lobes, as in the Labiatæ.   The
style is slender below, and thickest in the
upper part ; and the stigma is two-lobed.   The
leaves are opposite, and furnished with stipules

The flowers form a corymb in the Scarlet Verbena, and a spike in some of the other kinds, which elongates gradually as the flowers expand. The principal other genera are Clerodendron, or Volkameria, Vitex (the Chastetree), Lantana, Aloysia (the Lemon-scented Verbena), and Tectona (the Teak) which is so much used in the East Indies for ship-building.

## ORDER CXLIV.—MYOPORINÆ.

AUSTRALIAN and Polynesian plants, nearly allied to Verbenaceæ. The principal genera are Myoporum and Avicennia, the White Mangrove of Brazil.

## ORDER CXLV.—ACANTHACEÆ.

THESE plants are known by the elastic opening of the capsules, which are two-celled, and the hooked points of the seeds by which they are attached to the placenta. The calyx remains on the ripe fruit, but in most of the plants it is so extremely small as to be inconspicuous, and its place is supplied by three large leafy bracts. The corolla varies considerably, being sometimes two-lipped as in Justicia, sometimes funnel-shaped as in Ruellia,

and sometimes campanulate, with a spreading
five-cleft limb, as in Thunbergia. There are
only two stamens in Justicia and some of the
other genera, but in Thunbergia, Acanthus,
and Ruellia, there are four of unequal length,
inclosed within the throat of the corolla. The
ovary is imbedded in the disk, and it is two
or many seeded; the style is simple, and the
stigma one or two lobed.

### ORDER CXLVI.—OROBANCHEÆ.

Leafless parasites, with brown or colourless
scaly stems and flowers. The genera are La-
thræa and Orobanche.

### ORDER CXLVII.—LENTIBULARIÆ.

Pretty little marsh plants, natives of Europe and North America. Pinguicula has very
much the appearance of a violet, and the Utri-
cularias are floating plants.

### ORDER CXLVIII.—PRIMULACEÆ, THE
### PRIMROSE TRIBE.

The plants belonging to this order are easily
known by the stamens, or rather anthers, for

they have scarcely any filaments, being affixed to the corolla in the centre of the lobes, instead of being alternate to them, and by the capsule, though five or ten ribbed, being only one-celled, with a central placenta, to which the seeds are attached. The calyx remains on the ripe fruit. In the genus Primula (the Primrose), the calyx is tubular, and strongly marked with five distinct angles, which end in as many teeth ; and the corolla is salver-shaped, with a contraction in the tube, at the insertion of the stamens, the five segments of the limb being wedge-shaped and notched. The style is slender, and the stigma capitate. The capsule opens naturally by ten teeth, which curl back. The Cyclamen, or Sow-bread, one of the genera belonging to this order, has the lobes of the corolla bent back ; and when the flower falls, the peduncle coils up in a most curious manner, so as to bury the seed-vessel in the earth. These plants have tuberous roots, which are so acrid as only to be eaten by the wild-boars. The seed-vessel of the Pimpernel (*Anagallis*) resembles a round case with a lid, which may be taken off, when it displays a great number of seeds, so closely packed, that no room is lost. The principal other genera are the

American Cowslip (*Dodecatheon*), Bear's-ear
Sanicle (*Cortusa*), *Soldanella*, the Water Violet
(*Hootonia*), and Loosestrife (*Lysimachia*).

---

## ORDER CXLIX.—GLOBULARIÆ.

PRETTY alpine plants, with blue flowers.

---

## ORDER CL.—PLUMBAGINEÆ.

THIS order probably belongs to Monochlamydeæ.  The principal genera are Sea Lavender (*Statice*), remarkable for the coloured footstalks of the flowers; Thrift (*Armeria*); and
Leadwort (*Plumbago*).  The corolla in these
plants is either monopetalous, with the stamens
free from the corolla and growing from beneath
the pistil, or with five petals, to which the stamens are attached.  There are five styles and
five stigmas, but only a one-celled and one-
seeded ovary.  The fruit is thin and dry.  The
pedicels of all the species of the Sea Lavender,
particularly of *Statice arborea*, are often mistaken for the flowers.

# CHAPTER II.

PHANEROGAMOUS PLANTS—DICOTYLEDONEÆ.—
II. MONACHLAMYDEÆ.

In all the plants contained in this division,
the stamens and pistils have either no floral
covering, or only one ; and as, when this is the
case, the covering is called the calyx, the plants
in this division are said to have no corolla.
Some botanists think that the calyx and corolla
have become intermixed, so as to form only one
covering, which they call the perianth ; a word
applied to the calyx and corolla together.

---

## ORDER CLI.—PLANTAGINEÆ.

The weed called Plantain, or Rib-grass, is
well known to all persons who keep birds, as it
is a food that cage-birds are very fond of.　It
is conspicuous by its strongly-ribbed leaves,
which form a flat tuft on the ground, and by
the large arrow-shaped anthers of its four stamens, which hang on very slender filaments.
The flowers are arranged in dense spikes,
and are green and inconspicuous.

E E 2

### ORDER CLII.—NYCTAGINEÆ.

THE Marvel of Peru (*Mirabilis Jalapa*), and
the other species of that genus, are the only
ornamental plants belonging to this order.
The flowers consist of a coloured calyx, surrounded by a five-toothed involucre, which
greatly resembles a calyx. The true calyx is
funnel-shaped, with a spreading limb, the lobes
of which are plaited, and notched at the margin; and which, with the tubular part, form at
the base a globular swelling, which incloses the
ovary. The stamens grow from beneath the
pistil, adhering together at the base, so as to
form a kind of cup. The ovary contains only
one seed ; and the style is long and slender,
terminating in a capitate stigma, divided into a
number of tubercles or warts. The lower part
of the calyx remains on the ripe fruit, harden-
ing into a kind of shell.

### ORDER CLIII.—AMARANTHACEÆ.

THE flowers of the plants belonging to this
order are either in spikes, like Love-lies-bleeding
(*Amarantus caudatus*), in heads like the Globe
Amaranth (*Gomphrena globosa*), or in a singular
crest-like shape, like the Cock's-comb (*Celosia
cristata*). In all, the flowers have no corolla,

and only a very thin and dry calyx, which is
surrounded by hard, thin, dry bracts, of the
same colour, each ending in a long point.
There are generally five anthers, and two or
three styles, with pointed stigmas; but the
capsule contains only one cell and one seed; and
when ripe, it divides horizontally in the middle,
like the capsule of the Pimpernel.

---

### ORDER CLIV.—PHYTOLACEÆ.

HERBACEOUS plants and shrubs, with racemes
of red, white, or greenish flowers. Phytolacca
is the principal genus; and one species, the Virginian Poke (*Phytolacca decandra*) is remarkable
for being found wild in climates so different as
Spain and Portugal, the north of Africa, Jamaica, and North America. The flowers are
greenish, tinged with red, and they are followed by very dark purple berries, which are
said to have been formerly used for colouring
port wine, but the juice having medicinal qualities, their use in Portugal is now prohibited.
Rivina belongs to this order.

---

### ORDER CLV.—CHENOPODEÆ.

THE plants belonging to this order bear considerable resemblance to those included in the

order Amaranthaceæ, but their flowers are disposed in loose clusters without bracts, and all
their parts are fleshy ; while the flowers of the
Amaranthaceæ are disposed in dense spikes
with bracts, which, as well as the divisions of
the flowers, are quite hard and dry.   The stamens are five in number, and they are spread
out like those of the Nettle tribe; there are
two styles with hairy stigmas, and the capsule
resembles the Echinus, or Sea Urchin.   The
principal genera in this order are,—Spinach
(*Spinacea*), Red and White Beet (*Beta vulgaris*), Mangold Wurtzel (*B. altissima*), Chard
Beet (*B. cicla*), the Strawberry Blite (*Blitum*),
Fat-hen or Goosefoot (*Chenopodium*); Glass-
wort, the ashes of which are used in making
glass (*Salicornia*), Saltwort (*Salsola Kali,* or
*Soda*), from the ashes of which soda is prepared ;
and the Garden Orache (*Atriplex hortensis*).
The leaves of all the species are somewhat succulent and pulpy, and they are frequently
stained with brilliant colours.

ORDER CLVI.—BEGONIACEÆ.

THE only genus in this tribe is Begonia,
the plants belonging to which have pretty
flowers, and strongly-veined leaves, which are
crimson on the lower side, with one half smaller

than the other, and each furnished with a
pair of large stipules. The flowers are male
and female; the first consist of four sepals,
two of which are much longer than the others,
and a beard of anthers, with the filaments united
into one common stalk, and each anther containing two cells for pollen. The female flowers have
five sepals; the lower part is thick and fleshy,
having three unequal wings. This part becomes
the capsule, and it is furnished with three stigmas, each of which has two curiously-twisted
lobes. The capsule when ripe has three wings,
one much longer than the others; and it is in
three cells, each containing a central placenta
with a double row of seeds, which are covered
with a beautifully reticulated skin.

---

ORDER CLVII.—POLYGONEÆ.—THE BUCKWHEAT.

This order comprehends the Rhubarb (*Rhe-
um*), the Dock (*Rumex obtusifolius*), Sorrel (*R.
acetosa*), the Buckwheat (*Polygonum Fagopy-
rum*), Persicaria (*P. orientale*), Water-pepper
(*P. hydropiper*), and Knot-grass (*P. aviculare*).
The leaves of these plants either sheath the
stem with the base of their petioles, or are
furnished with ochreæ, that is, with stipules
which are joined together so as to form a kind
of purse or boot. The flowers are inconspicuous,

and the fruit is a triangular nut, retaining the
calyx till it is ripe. The genera Eriogonum,
Calligonum, and Kœnigia, formerly included in
this order, are now formed into another, called
Eriogoneæ.

---

### ORDER CLVIII.—LAURINEÆ.—THE SWEET-BAY TRIBE.

THESE plants are known by their anthers,
which are two or four celled, with the valves
curling upwards when ripe, like those of the
Berberry, and the filaments are furnished near
the base with two kidney-shaped glands. The
male and female flowers are distinct ; the former
have six, eight, or twelve stamens, and a calyx
of four or six divisions united at the base. The
female flowers have a one-celled and one-seeded
ovary, with a simple style, and an obtuse-
crested stigma ; and four or more abortive stamens, furnished with glands, but without anthers. The most interesting plants contained in
this order are,—the Sweet Bay (*Laurus nobilis*),
the Sassafras-tree (*L. Sassafras*, or *Sassafras
officinale*), the Cinnamon-tree (*L. Cinnamomum,*
or *Cinnamomum verum*, or *zeylanicum*); the
Camphor-tree (*L. camphora*, or *Camphora offi-
cinarum*); and the Alligator Pear (*L. Persea*,
or *Persea gratissima*). All the plants belonging
to this order are aromatic, either in the leaves,
bark, or fruit.

Two small orders, Illigereæ and Hemandia-
ceæ, containing Indian plants rarely met with
in England, are introduced here by some botanists.

---

### ORDER CLIX.—MYRISTICEÆ.

THE only interesting plant in this order is the
Nutmeg (*M. officinale*, or *moschata*).  In this
plant, the fruit is pear-shaped, and it consists of
a half-fleshy pericardium enclosing a jet-black
stone, encircled by a fleshy orange-red arillus,
which is the mace.  The nutmeg is the kernel
of the stone, and it is not taken out for sale
till it is sufficiently ripe to rattle when shaken.
The leaves are of a dark green above, and glaucous beneath ; and the flowers are white, with
the red pistil conspicuous in the centre.  The
tree is a native of Ceylon and the East Indian
Islands, and it requires a stove in England.

---

### ORDER CLX.—PROTEACEÆ.

THE principal genera are Protea, Banksia,
Dryandra, and Grevillea, all very singular
plants, the species of which, when one of each
genus has been seen, are easily recognised.
They are all natives of the Cape of Good Hope
and New Holland.

### ORDER CLXI.—THYMELÆÆ.

THIS order is well known from the Mezereon
and the Spurge Laurel, both common garden
shrubs belonging to the genus Daphne. The
berries of both are poisonous, and the bark
acrid. The flowers of the Mezereon (*D. Meze-
reum*) have a coloured calyx, which is tubular,
with a four-cleft limb (see *fig.* 143), which is
slightly hairy on the outer surface, and pitted

on the inner one. It is said that
this calyx will separate readily
into two, the inner part peeling
off like a lining : but I have
never been able to effect this

FIG. 143.—A FLOWER   without tearing the outer cover-
OF MEZEREON.    ing.    There are eight anthers,
with scarcely any filaments, affixed in two rows
to the throat of the corolla ; and an egg-shaped
ovary, with a tufted stigma without any style.
The fruit is a drupe, that is, formed like a
plum, with a fleshy pericardium, enclosing a
stone or nut, the kernel of which is the seed,
and which sometimes appears to be partially
enveloped in a sort of hairy bag, which is the
lining of the ovary become loose. The flowers
of the Mezereon grow round the stem, with a
tuft of leaves at the top ; but those of the
Spurge Laurel (*D. Laureola*) are in a cluster of

short drooping racemes. The most remarkable species of the genus is, however, the Lace Bark-tree of Jamaica (*D. Lagetto*, or *Lagetta lintea-ria*), the liber or inner bark of which has such tough fibres as to bear stretching out consider-ably without breaking ; and in this state it re-sembles lace so much, that a collar and ruffles were made of it and sent to Charles II. Gnidia, a greenhouse plant, has little scales in the mouth of the calyx ; and Pimelea has the flowers in heads, surrounded by a four-leaved involucre. The principal other genera are Lachnæa, a little Australian plant with woolly flowers, Passe-rina or Sparrow-wort, and Struthiola. The curious little tree called Leatherwood (*Dirca palustris*) also belongs to this order.

## ORDER CLXII.—OSYRIDEÆ.

Exotic trees with white or greenish flowers. The only genera are the Poet's Cassia (*Osyris*), and a genus of Australian plants called Exocar-pos.

## ORDER CLXIII.—SANTALACEÆ.

The most interesting plant is the Sandal-wood tree (*Santalum album*), which requires a stove in England ; but the North American trees be-longing to the genus Nyssa, including the Tupelo-

tree and the Ogechee Lime, are quite hardy.
The flowers are small and insignificant; and the
fruit is a drupe.

- - -

### ORDER CLXIV.—ELÆAGNEÆ.

THE three genera included in this order are
the Sea Buckthorn (*Hippophae*), the Oleaster
(*Elæagnus*), and the Shepherdia; all so easily
recognised by their silvery foliage, as to need no
particular description. The flowers are small
and inconspicuous.

- - -

### ORDER CLXV.—ASARINEÆ, OR ARISTOLOCHIEÆ.

THE genus Aristolochia, or Birthwort, is remarkable for the very singular shape of its
flowers, which are as strange, and as much varied,
as it is possible for the wildest imagination to
conceive. The flowers are tubular, with one lip
much longer than the other; and the tube takes
an abrupt bend near the middle. Here are six
anthers, fixed very curiously on the outside of a
club-shaped column, split into six lobes at the
point. In the centre of this column is a style
with a six-rayed stigma; and the fruit is a large
capsule with six cells, which opens by as many
slits, and discharges the numerous thin, flat,
dark brown seeds.

*Asarum canadense*, the Wild Ginger of North

America, has kidney-shaped leaves, and dark purplish brown flowers, on very short footstalks, which resemble those of the genus Stapelia, both in appearance and smell.

---

### ORDER CLXVI.—CYTINEÆ.

THE most interesting plant in this order is *Nepenthes distillatoria*, the Chinese Pitcher-plant, the leaves of which have a tendril at the point curiously dilated at the extremity, so as to form a cup-like appendage, which is generally full of water. The rim of the pitcher is beautifully ribbed, and it is furnished with a lid. The male and female flowers are on different plants, but neither of them possess much beauty. The remarkable Javanese fungus Rafflesia belongs to this order.

A small order called Cephaloteæ, and containing only the genus Cephalotis, formerly included in Rosaceæ, is introduced here.

---

### ORDER CLXVII.—EUPHORBIACEÆ.

THE genus Euphorbia is well known by the British weed called Caper Spurge, and the showy stove plants which belong to it. The male and female flowers are distinct; but both are inclosed in one cup-like involucre. In *fig.* 144, *a* is the

involucre, *b* the female flower, and *c* the male
ones. The fruit (*fig.* 145) consists of three

FIG. 144.—EUPHORBIA.        FIG. 145.—FRUIT OF
                            EUPHORBIA.

carpels, each containing a single seed, which
divide with elasticity when the seeds are ripe.
All the plants belonging to this order have a
milky, glutinous juice when young, which in
some genera becomes solid when exposed to
the air. This is particularly the case with
Siphonia Hevea, a Brazilian tree, the sap of
which yields the Indian rubber used for Mac-
intosh cloaks, &c.; it being more suitable for
that purpose than the caoutchouc yielded by
the *Ficus elastica*, which is the true Indian Rubber. The principal other genera belonging to
this order are the Box (*Buxus*), the tree kind
of which yields the wood used for wood-engrav-
ing, and the dwarf variety is employed as edging
for gardens ; Croton, an annual species of which
(*Croton Tiglium*) yields the celebrated Croton
oil; the Cassava (*Jatropha Manihot*), which
though poisonous in a raw state, becomes the
wholesome food called tapioca, when properly

prepared; Palma Christi (*Ricinus communis*), from the seeds of which castor-oil is made; and the Manchineel tree (*Hippomane*), which is said to be so poisonous as to occasion the death of those who sleep beneath its shade.

### ORDER CLXVIII.—STACKHOUSEÆ.

SMALL Australian shrubs with insignificant flowers.

### ORDER CLXIX.—ANTIDESMEÆ.

EAST-INDIAN trees with inconspicuous flowers.

### ORDER CLXX.—URTICEÆ.—(SEE CHAP. X. P. 157.)

THE plants belonging to this order are divided into two sections, viz. those with tough fibres, as the Hemp, the Nettle, &c.; and those with milky sap, such as the Fig, the Mulberry, the Bread-fruit, &c. All the genera have the male and female flowers separate. The male flowers have four stamens which spring back and discharge their pollen with elasticity, and the female flowers have a one-celled ovary with two long stigmas.

## ORDER CLXXI.—ULMACEÆ.

The principal genera are the Elm (*Ulmus*), the Nettle-tree (*Celtis*), and the Zelkoua-tree (*Planera*). The flowers, though very small, are pretty, from their opening in clusters before the leaves ; and each has four stamens, with dark purple anthers, and is furnished with dark brown bracts. The fruit is a utricle, having a single seed, encircled by a broad thin transparent membrane. The leaves are rough, and their sides are unequal at the base. The bark of Elm trees is rough and deeply furrowed ; and the roots spread, instead of penetrating deeply into the ground like those of the Oak.

## ORDER CLXXII.—PIPERACEÆ.

The species are generally climbing plants with perfect flowers, which are produced in spikes, and are succeeded by one-seeded berries. The genus Piper contains the common Pepper (*P. nigrum*), the Betel (*P. Betel*), and several other species.

## ORDER CLXXIII.—JUGLANDACEÆ.—(See Chap. XI. P. 176.)

The male and female flowers are distinct, but on the same plant. The male flowers are pro-

duced on long thick anthers, and each consists
of a scale-like calyx enclosing numerous stamens ;
the female flowers are two or more together,
and each consists of a scale-like calyx, enclosing a
one-celled ovary.  The fruit is a drupe, that is,
it consists of a fleshy husk enclosing a nut.
The embryo fills the whole seed ; and the cotyledons are fleshy, two-lobed, and wrinkled.
There are only two genera, the Walnut (*Juglans*), the male catkins of which are produced
singly, and the Hickory (*Carya*), the male
catkins of which are in clusters.

### ORDER CLXXIV.—AMENTACEÆ.—(See Chap. XI. P. 174).

The male flowers are in catkins, and the fruit
of most of the genera is, when ripe, partially
or wholly enclosed in a cup-like involucre,
formed by the adhesion of the numerous bracts.

### ORDER CLXXV.—HAMAMELIDEÆ.

This order has been already inserted, p. 380.

### ORDER CLXXVI.—EMPETREÆ.

Little heath-like plants, with small flowers
and showy berries.  The Crowberry, *Empetrum
nigrum*, is very common in Scotland on heaths.

ORDER CLXXVII.—CONIFERÆ. (See Chap. XII. P. 205.)

THE male and female flowers are both produced in catkins, and both consist only of scales.
The pollen of the male flowers is very abundant, and is discharged freely in fine weather.
The female flowers form cones, consisting of
numerous scales, at the base of each of which
are two winged seeds. The timber abounds in
resin.

ORDER CLXXVIII.—CYCADEÆ. (See Chap. XII. P. 229.)

THESE singular plants have thick timber-like
trunks, yet they can hardly be called trees, as
they increase in height by a single terminal
bud. The leaves are pinnate, and they unroll,
when they expand, like those of the ferns.
The male flowers are in cones, and the female
ones either in cones, or produced on the margin
of contracted leaves. The principal genera are
Zamia and Cycas, and one species of the latter
yields a kind of sago; the true kind being a
product of a species of Palm.

# CHAPTER III.

## PHANEROGAMOUS PLANTS.—MONOCOTYLEDONEÆ.

ALL the trees belonging to this division are natives of tropical countries ; and they, as well as all the herbaceous plants belonging to it, are distinguished by the veins of their leaves being never branched, but principally in parallel lines. These plants are re-divided into those with a perianth, which are called the Petaloideæ, and in which are included the Orchidaceæ and the bulbous-rooted plants ; and those without a perianth, which are called Glumaceæ, and in which are included the grasses, and sedges.

## § I.—PETALOIDEÆ.

### ORDER CLXXIX.—HYDROCHARIDEÆ.—THE FROG'S BIT TRIBE.

AQUATIC plants, two of which are of very curious construction. In Vallisneria, the male and female flowers are on different plants, and the buds of the female flowers rise on long spiral stalks, which gradually uncoil, till the flower appears above the surface of the water,

where it expands.  The male flowers are produced on separate plants at the bottom, but,
before they expand, they detach themselves
from the soil, and rise up to the surface, where
they float till the flowers have opened, and the
pollen has fallen on the stigmas of the female
flowers, after which the male flowers wither,
and the female ones coil up their stalks again
to ripen the seed-vessels at the bottom.  This
curious arrangement is necessary, because the
pollen should be dry when it falls on the stigmas ; and nearly a similar arrangement takes
place with the Fresh-water Soldier (*Stratiotes*).
The Frog's Bit (*Hydrocharis morsus ranæ*) is a
floating plant, with pretty white flowers.  *Da-
mosonium indicum* is a very handsome water-
plant, with white flowers and winged stems.

---

### ORDER CLXXX.—ALISMACEÆ.—THE WATER
### PLANTAIN TRIBE.

The principal genera in this order are Alisma,
Sagittaria, and Actinocarpus, all common
British aquatic plants.  The Water Plantain
(*Alisma plantago*) has ribbed leaves, and a
loose panicle of small pinkish flowers, which
have a permanent calyx of three sepals, a corolla
of three petals, six stamens, and numerous carpels, which grow close together so as to form a

head, as in the Ranunculus tribe. *A. natans*, which is generally found on lakes in the mountainous districts of Wales and Cumberland, has rather large white flowers, with a yellow spot at the base of each petal. The flower-stalks rise high above the water, and the flowers expand in the months of July and August. The common Arrowhead (*Sagittaria sagittifolia*) has curiously-shaped leaves, resembling the head of an arrow. The flowers are white, and resemble those of *A. natans*; but they have a pink spot at the base, and there are numerous stamens. The flowers are in whorls, and those in the upper whorls are generally destitute of carpels. The common Star-fruit (*Actinocarpus damsonium*) has only six carpels, which are so arranged as to form a star-like fruit when ripe.

---

ORDER CLXXXI.—BUTOMEÆ.— THE FLOWERING RUSH TRIBE.

THE flowering Rush (*Butomus umbellatus*) is certainly the handsomest of the British aquatic plants. The flowers are rose-coloured, crimson, or white; and they are produced in large erect umbels. The calyx and the corolla are generally of the same colour, and in three divisions each; there are nine stamens and six capsules, which are many-seeded. The leaves are trian-

gular or flat.  *Limnocharis Plumieri* is a very
handsome Brazilian aquatic belonging to this
order.

------

## ORDER CLXXXII.—JUNCAGINEÆ.—THE ARROW-GRASS TRIBE.

INSIGNIFICANT bog plants, with grassy leaves,
and central spikes or racemes of greenish yellow
flowers.

------

## ORDER CLXXXIII.—ORCHIDACEÆ.

THE plants belonging to this order may be
divided into two kinds, those that grow in the
earth, and those which require to have their
roots suspended in the air ; the latter being the
beautiful tropical plants called Orchideous Epiphytes.  Most of the terrestrial Orchidaceæ are
British plants belonging to several genera, the
most curious of which are Orchis, Habenaria,
Ophrys, Aceras, Nœttia, Epipactis, and Ma-
laxis.  Nearly all the British Orchidaceæ have
tuberous roots, which remain above ground,
a new tuber being formed every year.  The
leaves are alternate, with an entire margin,
without any footstalk, and sheathing the stem
at the base.  The flowers are produced in a
spike, furnished with bracts, and though they
are very irregular in their forms, there are cer-

tain particulars in which they all agree. Though in reality sessile, they appear to have each a footstalk, but this footstalk is only the long twisted ovary (*c* in *fig.* 146), which is one-

FIG. 146.—ORCHIS MORIO.

celled and many-seeded, and which serves to support the calyx and corolla of the flower, which are both above it. The calyx consists of three sepals, one of which has the appearance of a hood (*a*), and the others (*b b*) look like wings. The petals are very disproportionate in their size; two are generally very small, and are only seen peeping beneath the hood of the calyx; while the third (*d*), which is called the labellum, or lip, is very large, and hangs down. In the centre of the flower is a singular mass, called the column, composed of the stamens and pistil, grown together (see *a* in *fig.* 147). In this column there is one perfect

FIG. 147.—POLLEN MASSES OF THE ORCHIS.

anther (*b*), and two imperfect ones (*c c*). The perfect anther consists of a pouch or bag, which, when opened, displays two stalked masses of globular pollen, one of which (*d*) is pulled

down to show its appearance, while the other
remains in its case at (*e*). The stigma is a
sort of cup half full of a glutinous fluid, but it
appears entirely shut out from the pollen,
which is not only enclosed in its pouch or bag,
but is of such a solid waxy nature as to prevent
any possibility of its being carried by wind or
insects to the stigma. Nature, however, has
contrived a means of obviating the difficulty.
At the foot of each stalk of the pollen masses,
there is a little protuberance, covering a gland,
through which the pollen descends to the stigma, and thence to the ovary or germen.

The different genera are distinguished, partly
by the manner in which the granules of the
pollen adhere together, and partly by the shape
of the flowers; and their different species vary
principally in the form of the labellum. In the
genera Orchis and Habenaria, the labellum is
drawn out behind into a kind of spur (see *e* in
*fig.* 146); and in others it assumes strange
shapes, as in the Man Orchis (*Aceras anthropo-
phora*), where the labellum looks like a little
man; and in the Lizard Orchis (*A.* or *Orchis
hircina*) where the labellum is drawn out into a
long tail, which looks like the tail and long
body of the lizard, while the petals, which are
long and narrow and bent back, look like the
hind legs. In the genus Ophrys, the labellum

also takes strange shapes, sometimes resembling a bee, at others a fly, and at others a spider. In the genus Cypripedium, the two side stamens bear anthers and pollen, and only the central one is imperfect.

In the orchideous epiphytes the same general construction prevails, but the forms of the flowers are still more varied and fantastic. All of them have pseudo bulbs above ground, which serve as substitutes for the tubers of the terrestrial species.

---

### ORDER CLXXXIV.—SCITAMINEÆ.

This order contains several plants, well known for their useful properties, as for example, the Ginger (*Zingiber officinale*), and the Turmeric (*Curcuma Zerumbet*). Some of the plants grow tall and reed-like, as for example in Hedychium. Most of the genera have a creeping underground stem, called a rhizoma, which is often jointed. The flowers are produced in spathe like bracts ; the calyx is tubular, and adheres to the ovary ; and the corolla, which is also tubular, has six segments arranged in two rows ; the inner row, which is supposed to consist of the dilated filaments of abortive stamens, has one of the segments, called the labellum, larger than the rest. There are three stamens, two

of which are abortive, as in the Orchidaceæ ;
but the pollen does not cohere in masses, and it
is not inclosed in a kind of pouch or bag.   The
ovary is three-celled (though the cells are sometimes imperfect), and many-seeded ; the style
is filiform, and the stigma is dilated and hollow.
The fruit is generally a capsule ; but in some
cases it is a berry.

## ORDER CLXXXV.—CANNEÆ.

THE most interesting genera are—*Canna*, containing reed-like plants with brilliant flowers ;
as, for example, *C. indica*, the Indian Shot ;
*Thalia*, a curious aquatic ; and *Maranta*, the
tubers of which furnish India Arrowroot.   The
flowers in their construction greatly resemble
those of the preceding order ; but the filaments of the stamens are petal-like, and it is
one of the side stamens that is perfect, the
middle and the other side stamens being always
abortive.   The fruit is always capsular.

## ORDER CLXXXVI.—MUSACEÆ.

THE genus Musa is known by its fruit, which
is eaten under the names of Plantain and Ba-
nana.   The flowers are produced in spikes, enclosed in spathe-like bracts, which are often

richly coloured ; and they are succeeded by the
fruit, which hang down in massive spikes of
enormous weight.    The leaves are very large
and strong, and Indian muslin is manufactured
from the fibres of one of the species.    The principal other genera, Strelitzia and Heliconia, are
both remarkable for the brilliant colours of
their flowers.

ORDER CLXXXVII.—IRIDACEÆ.

The principal genera belonging to this order
are—Iris, Moræa, Marica, Vieusseuxia, Homeria,
Sisyrinchium, Patersonia, Witsenia, Ferraria,
Tigridia, Babiana, Watsonia, Gladiolus, Spa-
raxis, Tritonia, Ixia, and Crocus; but almost
every genus contained in the order has showy
flowers, and is consequently well known in gardens.    The leaves are generally thin, long, and
flat, with the edge towards the stem, and the
flowers are produced from spathes ; the perianth
is also in six segments coloured alike, the calyx
and corolla being in most cases confounded together.    The genus Iris has generally tuberous or
solid bulbous roots, of the kind called corms, and
the perianth of the flower is divided into six
segments, three of which are larger than the
others ; these three larger segments, which form

the calyx, (see *a* in *fig.* 148) are reflexed, and
a stamen springs
from the base of
each, which re-
clines upon it,
with its anther
turned from the
rest of the flower,
the segment, in
many species,
having a kind of
crest or beard
near the base, as
though it were in-

FIG. 148.—FLOWER OF THE IRIS.

tended to form a cushion for the stamen to
repose on, while over each stamen is spread, as
a kind of coverlid, a stigma (*b*) which is dilated
so as to resemble a petal. The petals (*c*) often
stand erect, and were called by Linnæus the
standards. The seed-vessel, which forms below
the flower, is a three-celled capsule, opening,
when ripe, by three valves, and containing
numerous seeds.

The other genera differ from the Iris in having
the lower part of the segments of the perianth
generally combined into a tube, with the ovary
below, looking like a footstalk ; the limb being
divided into six parts, all so much alike, both
in form and position, as to render it difficult to

distinguish the calyx from the corolla. There is only one style, with three stigmas, which are always more or less leafy, and the anthers (which are never more than three) are always turned away from the pistil. In Ferraria, the filaments of the stamens grow together, and form a hollow tube, as in the Passion-flower, surrounding the style and stigmas; and in the Saffron Crocus (*C. sativa*), the stigmas (which, when dried, form the saffron) are so heavy, as to hang out on one side of the perianth from between the segments. Most of the genera have solid bulbs or corms at the base of their stems; but some, such as Marica, Sisyrinchium, and Patersonia, have only fibrous roots. The genera Colchicum and Bulbocodium very much resemble the Crocus in the appearance of their flowers; but they are distinguished by having three styles and a superior ovary, and they are included in the order Melanthaceæ.

---

## ORDER CLXXXVIII.—HÆMODORACEÆ.

THE principal genera are Wachendorfia, Hæmadorum, and Anigozanthos, which differ from the preceding genus principally in having six stamens, the anthers of which are turned towards the stigma. Most of the genera have only fibrous roots, but Wachendorfia has a rhizoma, producing buds in the scales. The plants are natives of the Cape of Good Hope and New Holland, and the roots of some of the species yield a brilliant scarlet dye.

## ORDER CLXXXIX.—HYPOXIDEÆ.

BULBOUS-ROOTED plants, with long narrow leaves covered with soft downy hairs, and rather small yellow flowers, which are frequently fragrant.

## ORDER CXC.—AMARYLLIDACEÆ.

A LARGE order of genera, all of which have bulbous roots, and most of them splendid flowers. Some of the most interesting genera are—Amaryllis, Nerine (the Guernsey Lily), Brunsvigia, Hæmanthus, Crinum, Pancratium, Narcissus, Galanthus (the Snowdrop), Leucojum, Alstrœmeria, and Doryanthes. The different kinds of Amaryllis have large lily-like flowers, divided into six equal segments, which are joined into a tube below, with six stamens, the anthers of which are turned towards the pistil, and a long style crowned with a simple stigma. The ovary is beneath the other parts of the flower, to which it serves as a receptacle; and in most of the plants it looks like a small green calyx below the perianth. The leaves are very long, but they are rather thick and fleshy, and their edge is not turned towards the stem. In Narcissus, Pancratium, and some other genera, the flowers have a kind of cup within the perianth, formed of the filaments of abortive stamens grown together. In Pancratium, the filaments of the anther-bearing stamens grow into the others, so as to form a part of the cup, the anthers springing from the margin of it; but in Narcissus, the fertile stamens are distinct. In Galanthus, and its allied genera, the anthers open by pores, as in the Ericaceæ, and there is a kind of receptacle on the germen, in which the petals, and sepals, and the filaments of the stamens, are inserted.

---

### ORDER CXCI.—HEMEROCALLIDEÆ.

THIS order, which included the Day Lilies (*Hemerocallis* and *Funkia*), the African Lily (*Agapanthus*), the Aloe (*Aloë*), the Tuberose (*Polianthes*), with several other genera which have their flowers in upright racemes or umbels, is now generally considered to form a section of the order Liliaceæ.

---

### ORDER CXCII.—DIOSCOREÆ.

THE Yam (*Dioscoreæ*), and the Elephant's-foot (*Testudinaria*), are the principal genera in

this order ; and both have an enormously-large
tuberous root which is eatable, and a very
slender climbing stem, with rather small leaves
and inconspicuous flowers. The ovary is below
the flower, and the fruit is capsular.

---

ORDER CXCIII.—TAMACEÆ.

THIS order consists only of the genus *Tamus*,
the Black Bryony, which Dr. Lindley includes
in Dioscoreæ. It has, however, a berry-like
fruit.

---

ORDER CXCIV.—SMILACEÆ.

THIS order includes Smilax, the root of a
species of which affords the drug called Sarsa-
parilla, the Lily of the Valley (*Convallaria*),
and the Alexandrian Laurel, or Butcher's
Broom (*Ruscus*). The male and female flowers
in Smilax are on different plants ; and in Ruscus the flowers spring from the middle of the
leaves. The perianth is in six equal segments,
and there are six stamens. The ovary is three-
celled, with the cells one or many seeded,
and the fruit is a globose berry. The seeds,
when ripe, have a brown membranous skin.
Dr. Lindley confines this tribe to Smilax, and
Ripogonum ; and includes the other genera in
Liliaceæ.

## ORDER CXCV.—ASPHODELEÆ.

THIS order includes the Hyacinth (*Hyacinthus*), the squills (*Scilla*), the onions (*Allium*), the Grape Hyacinth (*Muscari*), the Star of Bethlehem (*Ornithogalum*), King's Spear (*Asphodelus*), *Anthericum*, *Albuca*, *Gagea*, *Thysanotus*, *Asparagus*, the Dragon-wood (*Dracæna*), and New Zealand flax (*Phormium*). Many of these plants have tunicated bulbs; that is, bulbs which consist of several fleshy tunics or coats, which may easily be separated from each other, as may be seen in the hyacinth and the onion. The leaves are fleshy, and ligulate or strap-shaped; and the stems are frequently hollow. The flowers are generally in upright racemes, or umbels; they are regular, and sometimes bell-shaped; the perianth is divided into six segments, which are sometimes partly united into a tube, and recurved at the tip. There are six stamens attached to the perianth, and the fruit is either a fleshy or dry three-celled capsule, generally with several seeds, and opening into three valves, when ripe. Dr. Lindley makes this a separate order in his *Ladies' Botany*, but he combines it with Liliaceæ in his *Introduction to the Nat. Syst.*, and Sir W. J. Hooker includes in it Yucca and Aloe, the first of which in the *Hortus Britannicus* is included in Tulipaceæ, and the latter in Hemerocallideæ.

### ORDER CXCVI.—TULIPACEÆ.

THIS order in the *Hortus Britannicus* comprises the genera Yucca, Tulipa, Fritillaria,
Cyclobothra, Calochortus, Lilium, Gloriosa,
and Erythronium (the Dog Violet); but Sir
W. J. Hooker omits Yucca, and adds Bland-
fordia, Hemerocallis, and Polianthes; while
Dr. Lindley includes all these plants, together
with those comprised in Asphodeleæ, in the order
Liliaceæ. This last appears the most natural
arrangement, as all these plants have a regular
perianth of six segments, with six stamens,
and a dry or fleshy capsule of three cells, opening by as many valves. Some of the genera
have more seeds than others, and some of the
seeds have a hard, dry, black skin, while others
have the skin spongy and soft. Some of the
genera have the flowers erect and single, as in
the Tulip; in others the flowers are erect, but
in umbels, as in the Orange Lily; and in
others they are in racemes and drooping, as in
the Yucca, or single and drooping, as in the
Fritillaria, or with the segments curved back
as in the Martagon Lily.

## ORDER CXCVII.—MELANTHACEÆ.

THE plants belonging to this order have generally inconspicuous flowers, except Colchicum
and Bulbocodium, both of which have flowers
like the Crocus.   The bulbs of the Colchicum
are used in medicine ; but they and the whole
plant abound in an acrid juice, which is poisonous if taken in too large a dose.   The root
of Veratrum is also poisonous, and this plant
is believed to be the Hellebore of the ancients.
The Colchicum and the Bulbocodium are distinguished from the Crocus genus, which they
so strongly resemble in the appearance of their
flowers, by the ovary being within the flower
instead of below it, as is the case with all the
Amaryllidaceæ, and by their having three distinct styles, instead of one style and three
stigmas. In all other respects they are the same.

## ORDER CXCVIII.—BROMELIACEÆ.

THIS order includes the Pine Apple (*Bro-
melia Ananas*), the American Aloe (*Agave americana*), *Billbergia*, the magnificent plant *Bo-
napartea juncea*, now called *Lyttæa geminiflora*,
and the curious epiphyte *Tillandsia*.   What we
are accustomed to call the fruit of the Pine

Apple is, in fact, a fleshy receptacle, like that of the Strawberry, covered with scaly bracts, which are the remains of the fallen flowers. The flowers are blue, and one is produced in each bract; when they fall, the bracts thicken and grow together, and cover the ovaries, which sink into the fleshy part of the receptacle.

## ORDER CXCIX.—PONTEDERACEÆ.

ELEGANT aquatic plants, with kidney-shaped leaves, and spikes or racemes of blue or white flowers. The principal genus is Pontederia.

## ORDER CC.—COMMELINEÆ.

THIS order is principally known in Britain by the Spiderwort (*Tradescantia*), and the beautiful *Commelina cælestis*. Both plants have the flowers springing from a tuft of leaves which sheath the stem.

## ORDER CCI.—PALMÆ.—THE PALM TRIBE.

THIS order contains many lofty trees, which are, with one exception, without branches, and bear a tuft of large leaves, called fronds, at the summit. The flowers are small, with bracts, and they are enclosed in a spathe,

which bursts on the under side. The mass of
flowers is called a spadix; and it is succeeded
by the fruit, which, when ripe, is either a
drupe or a berry. In the Cocoa-nut Palm
(*Cocos nucifera*) the fruit is a drupe; but the
pericardium consists of hard, dry, fibrous matter, which is uneatable, the only part fit for
food being the albumen of the kernel. The
Date Palm (*Phœnix dactylifera*), and the Sago
Palm (*Sagus Rumphii*), are two interesting
plants, from their products.

## ORDER CCII.—PANDANEÆ.

THE most interesting plant in this order is
the Screw Pine (*Pandanus*), which has the
habit of the Palms, but the flowers of the
Arum tribe.

## ORDER CCIII.—TYPHINEÆ.—THE BULLRUSH TRIBE.

THE Bullrushes (*Typha*), also called Cat's-tail
and Reed-mare, are tall rush-like plants, with a
cylindrical mass of dark brown flowers round
the stem, surmounted by a spike of yellow
flowers. The lower dark-brown flowers are female ones, and the yellow ones are the males;
the former consist only of an ovary on a long
stalk, and a calyx cut into fine hairs so as to
form a kind of pappus. The male flowers have a

chaff-like calyx, enclosing the stamens, the filaments of which are united at the base, and the
anthers are very long and of a bright yellow.
The seed is a dry capsule, and the plant has a
rhizoma or creeping stem under the water.

---

### ORDER CCIV.—AROIDEÆ.—THE ARUM TRIBE.

THESE curious plants have their flowers in a
spadix, enclosed in a spathe, the male and
female flowers being separate, and the former
above the latter, with some abortive ovaries
again above them. The male flowers have
only one stamen in each without any covering;
and the female flowers in like manner consist
each of a single ovary, with a puckered-up hole
in the upper part, which serves as the stigma.
The fruit consists of a cluster of red berries,
which form round the spadix. Many of these
plants have a very unpleasant smell, and some
of them have a tuberous root, which, when
cooked, is eaten, though it is poisonous when
raw. *Arum* or *Caladium esculentum* is thus eaten
as a common article of food in the East Indies;
but the Dumb Cane (*A.* or *C. seguinum*) has its
English name from its juice being so poisonous
as, if tasted, to cause the lips to swell so as to
prevent the possibility of speaking. The beautiful marsh plant called *Calla* or *Richardia ethio-*

*pica*, or the White Arum, belongs to this order ;
as does the fragrant rush, *Acorus Calamus.* The
order Typhaceæ is included by many botanists
in Aroideæ ; and indeed, the difference between
them consists principally in the Bullrushes
having no spathe.

---

ORDER CCV.—FLUVIALES, OR NAIADES.—THE
POND-WEED TRIBE.

Floating plants, of which *Aponogeton dista-
chyon* is by far the most beautiful. This plant,
which is a native of the Cape of Good Hope,
has oblong, deeply-ribbed leaves on very long
footstalks, and the flowers in two-cleft spikes,
with snow-white bracts, which are very ornamental and very fragrant ; each flower consists
of from six to twelve stamens, and from two to
five carpels. The root is tuberous, and eatable when roasted. The Duckweed (*Lemna*),
which is sometimes included in this order, appears to consist entirely of a few leaves floating
on the water, each of which sends down a root ;
and many people believe that it never flowers.
If, however, it be watched in the months of
June and July, two yellow anthers will be seen
peeping out of the side of each leaf; and if the
opening be enlarged, the flower will be found to
consist of a kind of bag, open on one side, and
containing two stamens, with an ovary furnished

with a style and simple stigma. The fruit is a
one-celled capsule, containing one or more seeds.
Some botanists place this plant in a separate
order, called Pistiaceæ, from another genus included in it.

### ORDER CCVI.—JUNCEÆ.—THE RUSH TRIBE.

The most interesting genus is the Rush
(*Juncus*). These plants, low as they rank in
the vegetable world, have a regular perianth of
six divisions with six stamens, and a three-celled
capsule which opens by three valves. The
perianth of the flowers is, however, so small as
to be inconspicuous. Most of the species are
weeds, which are considered to indicate cold,
wet, and poor ground.

### ORDER CCVII.—GILLESIEÆ.

A grass-like plant, a native of Chili, with
greenish flowers.

### ORDER CCVIII.—RESTIACEÆ.—THE PIPEWORT TRIBE.

Rigid, inelegant, and often leafless, plants,
with the habit of rushes, natives of New Holland
and the Cape of Good Hope.

§ II.—GLUMACEÆ.

These plants, instead of having a regular
calyx and corolla, have nothing but green and
brown scales, which are called glumes, to cover
the stamens and pistil. There are only two
orders belonging to this division in British
fields and gardens.

----

ORDER CCIX.—CYPERACEÆ.—THE SEDGE TRIBE.

THESE plants have solid stems, and the leaves
not only sheathe the stem, but grow together
round it, so as to form a kind of tube. The
flowers are arranged in heads, some of which
contain only male flowers, each of which consists of a membranous scale and three stamens,
and others contain only female flowers. In the
genus Carex, the Sedge, these flowers are each
enclosed in a kind of bottle formed by two
scales growing together, and opening at the
top into two parts so as to show three stigmas,
which have only a single style. The fruit is
a dry, hard, triangular capsule with only one
seed. The most remarkable genera are *Papyrus*,
the plant anciently used for paper; *Scirpus*,
the Club-rush, used for making the seats of
chairs, mats, &c.; *Eriophorum*, Cotton-grass;
and *Cyperus*.

ORDER CCX.—GRAMINEÆ.—THE GRASS TRIBE.

THIS very important order includes not only
the common Grasses, but the Bread Corns, or
Cereal Grasses—Wheat, Oats, Barley, Rye, and
Maize; and the Sugar-cane and Rice. All these
plants are botanically allied to the Sedges, but
their stems are hollow, except at the joints,
where they become solid; and their leaves,
though sheathing the stem, do not unite round
it. The flowers are produced in spikes, which
are what are called spikelets. The glume, or
calyx as it was called by Linnæus, is generally
two-valved; and within it are two thinner
smaller scales, or paleæ, which were called the
corolla by Linnæus. Besides these, there are
frequently two still smaller scales within the
paleæ. There are generally three or six stamens, the anthers of which are two-celled, and
forked at the extremity. There are two styles,
either quite distinct, or combined at the base,
and the stigmas are feathery. The pericarp is
membranaceous, and adheres to the seed, forming a kind of caryopsides. The seeds contain a
great deal of albumen, which, when ground into
flour, becomes nourishing food. The stems, or
culms, are hollow and articulated; the leaves,
which are alternate, springing from each joint.
The most important genera are Wheat (*Tri-*

*ticum*), Barley (*Hordeum*), Rye (*Secale*), Oats
(*Avena*), Maize (*Zea*), the Sugar-cane (*Saccharum*), Rice (*Oryza*), and the Bamboo (*Bambusa*).
Oats are not produced in spikes, but in loose
panicles ; and the male and female flowers of
the Maize or Indian Corn are on different
plants.

# CHAPTER IV.

### CRYPTOGAMOUS PLANTS.

THESE plants are generally described as being without spiral vessels, and consisting only of cellular tissue; but spiral vessels are known to exist in the Ferns, and are said to have been found in the Mosses. Whether this be the case or not, it is evident that the plants included in this division are very different from all that have preceded them, and occupy a lower grade in the scale of vegetable creation. They are divided into two sub-classes: viz. the *Foliaceæ*, or those with leaves, and the *Aphyllæ*, or those without leaves; both of which are without visible flowers, though some have what are called anthers, and the Mosses have something resembling a style and stigma. They may also be said to have no seeds, for the spores, or sporules as they are called, are very different from the seeds of vascular plants, and they have neither cotyledon nor embryo.

## SUB-CLASS 1. FOLIACEÆ.

### ORDER CCXI.—FILICES.—THE FERN TRIBE.

THOUGH some of the Ferns are so common that almost every one must have seen them, very few persons are aware how very curiously they are constructed.   In the first place, they may be said to have neither stems nor leaves, and neither flowers nor seeds.   The different parts of the plant spring from a rhizoma, and the leaves, which are called fronds, have their veins neither branched nor in parallel lines, but forked.   On the back of the leaves are some curious brown spots of various shapes called sori; and these, which generally form under the outer skin or cuticle of the leaf, and which always spring from one of the veins, contain a number of small grains, called the thecæ, which are in reality cases containing the sporules or seeds.   When the sorus forms under the cuticle of the leaf, the membranous part raised, which resembles a blister, is called the indusium ; but sometimes the sori are naked, that is, they are formed on the outside of the cuticle ; and sometimes they are found on the margin of the leaf, which folds over them, and supplies the place of the indusium.   The order is generally divided into two sections, called Polypodiaceæ and Osmundaceæ.   The first contains those plants

which unroll their leaves, when they rise from
the stem, and which have their sori either on
the back or on the margin of the frond.  The
thecæ are on stalks, and they are furnished with
a ribbed, elastic, articulated but incomplete ring,
which seems to serve as a sort of hinge when
they burst.  This elastic ring is a continuation
of the stalk of the theca, which always bursts on
the opposite side.  The following are the principal genera in this division : Polypody (*Polypo-
dium*), sori without any indusium ; Shield Fern
(*Aspidium*), Bladder Fern (*Cistopteris*), and
Spleenwort (*Asplenium*), all of which have their
fronds pinnate or pinnatifid ; Maiden Hair
(*Adiantum*), Hart's-tongue (*Scolopendrium*), the
frond of which is simple and shaped like a tongue,
and the sori oblong ; and Brake (*Pteris*), the
leaves of which are pinnatifid, with the sori
placed round the margin so as to form a con-
tinuous line, and the edge of the leaf turned over
them.  The rhizoma of the Brake is eaten in
many countries, and the fronds, when burnt,
yield alkali, which is used in making both soap
and glass.

The second division Osmundaceæ comprises
those Ferns which apparently have flowers ; the
flowers, however, being merely sori, with the
leaves on which they grew shrivelled up round
them.  The most remarkable of these is the

flowering Fern (*Osmunda regalis*); but others
are—the Grape Fern or Moonwort (*Botrychium*),
a species of which, a native of North America, is
called there the Rattle-snake Fern; and the
Adder's Tongue (*Ophiglossum*). The Tree Ferns
of New Zealand are magnificent plants. The
trunk or stipe rises to the height of forty or
fifty feet without a branch, and then terminates
in a head of noble fronds, which hang down on
every side like a plume of feathers. The wood
of these trees when cut across, instead of being
in circles like the wood of Dicotyledonous trees,
or full of pores like that of the Endogens, is
marked with a number of zigzag lines, the traces
of the stalks of old fronds which have grown
together and formed the stipe.

ORDER CCXII.—LYCOPODINEÆ.—THE CLUB-MOSS
TRIBE.

These plants appear to occupy the intermediate space between the Ferns and the Mosses.
They have creeping stems, and grow two or
three feet high; the erect stems being clothed
with imbricated leaves, in the axils of which
these are produced. Some of them open into
three or four valves, and contain sporules;
while others are only two-valved, and contain a
kind of powder, which some suppose to be

pollen, and others abortive sporules. In some
of the species, the thecæ are produced in
bracteated spikes, which resemble the young
strobiles on a Spruce Fir. The seeds of the
common Club-moss (*Lycopodium clavatum*) are
used at the theatres to imitate lightning.

---

### ORDER CCXIII.—MARSILEACEÆ.

THESE are aquatic herbs, the thecæ or receptacles of which are always found in the axils of
the leaves near the root. In the genus *Isoetes*
(Quillwort) these are of two kinds, like those of
the Club-mosses, the one containing powder,
and the other granules; but in Pepper-grass or
Pill-wort (*Pillularia*), the receptacles are four-
celled, and each cell contains both powder and
granules. *Marsilea*, from which the order takes
its name, is a native of Italy and other parts of
the south of Europe, where it grows in the same
manner as Duckweed does with us.

---

### ORDER CCXIV.—EQUISETACEÆ.—THE HORSE-TAIL TRIBE.

THE thecæ of these well-known plants are
contained in terminal cone-like spikes or catkins, from four to eight lying in each scale.

The stems are tubular, and articulated with whorls of membranaceous sheaths, and of slender branches, jointed, and sheathed like the stem at every joint. All the species of Equisetum abound in silicious matter, and particularly the Dutch Rush (*E. hyemale*), which is used for polishing both wood and metal. The handsomest species is *E. sylvaticum*.

---

## ORDER CCXV.—CHARACEÆ.

AQUATIC herbs, contained in the genera Nitella and Chara, always growing under water, with slender jointed stems, surrounded at the joints by whorls of tubular leaves or branches, which are either membranaceous and transparent, as in *Nitella*; or brittle, and more or less encrusted with carbonate of lime, as in *Chara*, Stonewort. The organs of reproduction are formed in the axils of the branches, and consist of transparent globules, and hard, spiral nuculas, which appear to be formed of twisted leaves, the points of which often form a kind of crest. Young plants are only produced by the nuculas.

### ORDER CCXVI.—MUSCI.—THE MOSS TRIBE.

THE Mosses have fibrous roots, and slender
wiry stems, densely covered with leaves, which
are very small, and laid over each other like
scales (see *a* in *fig.* 149). The theca (*g*) is urn-

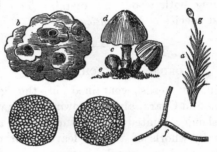

FIG. 149.—CRYPTOGAMOUS PLANTS.

shaped, and it is produced singly ; in most cases,
on a long, slender, wiry stem, called a seta,
which signifies a bristle, but sometimes without
any stalk. It always springs from a tuft of
leaves, differing both in size and shape from
ordinary leaves, which form what is sometimes
called the perichætium. Among these may
occasionally be seen a few stalks, resembling the
Lichen called Cup-moss, which terminates in a
kind of cup, and thickened at the base. The
cups and upper parts soon die away, and the
thicker part left among the leaves swells, and

in time rises on a stalk of its own, carrying away one of the leaves with it on its head. This is the theca, and the leaf it carried away, and which resembles an extinguisher, is called the calyptra, and it remains on till the sporules are nearly ripe. When the calyptra falls, the theca is found to be covered with a little lid called the operculum; which also falls off in time, and shows the mouth or stoma of the theca. This mouth is sometimes naked, and sometimes covered with a kind of film; but generally it is surrounded by a row of long, slender, hair-like teeth called the peristome or fringe. When there are two rows of these hair-like teeth, the inner ones, which are finer than the others, are called the cilia; and the number of both the cilia and the teeth is always some number that can be divided by four. In the cavity of the theca is a central axis called the columella, and around that are found the sporules, kept together by the lining of the theca, which forms a kind of open bag. This is the usual construction of all the numerous genera of mosses; but in some kinds, as for example in the Hair-moss (*Polytrichium*), in addition to the theca, a number of granules are found among the leaves, which are said to be capable of producing young plants.

## ORDER CCXVII.—HEPATICÆ.

THESE plants greatly resemble Mosses in their appearance, but they differ in their construction. The theca has no lid, but bursts into valves; and it generally contains not only sporules, but tubes formed of curiously twisted threads, called elaters. Jungermannia and Marchantia have a calyptra, which the other genera are without; and in Jungermannia the theca has a sort of sheath, which is sometimes called the calyx. There are also stalked granules called anthers, and warts which form on the leaves, and break up into a kind of sporules.

## SUBCLASS II.—APHYLLEÆ.

### ORDER CCXVIII.—LICHENES.

THOUGH these plants are said to have no leaves, they consist almost entirely of a kind of leafy stem, called a frond or thallus, the branches of which are called podetia (see *a* in *figs.* 150, 151, and 152). The spores or sporules are produced in what are called shields (*b* in *figs.* 149, 150, and 151), which are generally embedded in the thallus, and which, when they

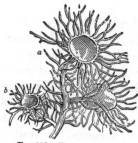

FIG. 150.—USNEA FLORIDA.
(Old Trees.)

are cup-shaped (as in *fig.* 150), are called scyphæ, and when flat (as in *fig.* 151), apothecia. The sporules, which are very numerous, are inclosed in receptacles of various forms, which are embedded in the shields.

Some of the commonest lichens are *Usnea florida* (*fig.* 150), and *Ramalina fastigiata* (*fig.* 151), both of which are found on old oaks, and are generally called grey moss; and *Cornicularia hetero-*

FIG. 151.—RAMALINA FASTIGIATA.
(Rocks and Trees.)

*malla* (*fig.* 152) is a brown mossy-looking lichen,

FIG. 152.
CORNICULARIA HETERO-
MALLA.—(Old Trees.)

often found on the bark. Other more interesting lichens are—the Iceland-moss (*Cetraria islandica*), the Reindeer-moss (*Cenomyce,* or *Cladonia rangiferina*), the Cup-moss (*Cenomyce pyxidata*), and the Orchil (*Rocella tinctoria*).

## ORDER CCXIX.—FUNGI.

THE Fungi are divided into several distinct
sections ; the most important of which may be
called the Mushroom tribe. The largest genus
in this division is Agaricus, and the plants belonging to it consist of a stipe, or stalk (*c* in
*fig.* 149), surmounted by the pileus or cap (*d*).
When the mushroom first appears, the stalk is
covered by a thin membrane, called the veil (*e*),
which unites the cap to the lower part ; but as
the mushroom grows, this veil is rent asunder,
and it either entirely disappears, or only a small
part of it remains round the stalk, which is
called the annulus or ring. Under the cap are
the gills or lamellæ, which are of a dark reddish
brown ; and attached to these are the thecæ,
containing the sporules or seed. In the common Mushroom (*Agaricus campestre*), and all the
eatable kinds, the gills are pink when the veil
breaks, which it does very soon, and they become afterwards nearly black ; but in all the
poisonous kinds, the veil is longer before it
breaks, and when it does so, the gills are pale,
and frequently nearly white, without becoming
darker ; the smell is also quite different. The
Mushroom tribe, which includes all the Fungi
that carry their sporules in the part above
the stem, is divided into two sections, viz.,

those with caps, like the Mushroom, and those which are slender and entire, but club-shaped in the upper part, like *Clavaria helvola*, a fungus often found in meadows, which resembles the stamen of an orange-lily.

The Morel tribe includes those Fungi which have their sporules in the stipe, and it is in two divisions; the first of which includes those which, like the Morel (*Morchella esculenta*), have a pileus, or cap, like a mitre; and the second, those which have the pileus curving upwards, like a cup, as in Peziza. A third tribe includes those which, like Tremella, are of a jelly-like substance; and in a similar manner all the numerous genera are arranged. Among these the most remarkable are the Truffle (*Tuber cibarium*), which is found buried in the earth, and the curious Fungi called Blight and Mildew, which belong to several different genera, and which appear on the leaves and fruit of other plants.

## ORDER CCXX.—ALGÆ.

THE Sea-weeds are placed on the extreme verge of the vegetable kingdom; and indeed some of them seem almost to partake of the nature of zoophytes. They can live only where there is abundance of moisture, and many of

them, such as the different kinds of Fucus,
inhabit the sea; by the waves of which they
are torn up from their native beds, and washed
on shore by the tides.  Others are found in the
form of Confervæ, or green slime, on the surface of stagnant ponds, or on damp stone or
gravel-walks; and others appear to form one
of the connecting links between vegetable and
animal life, as the joints in which they are produced possess the power of separating from
each other, and in their divided state so closely
resemble animals, as to puzzle naturalists to
know where to place them.  The Algæ are
divided by botanists into three classes; viz.,
the jointless, the jointed, and the disjointed.
The jointless Algæ are by far the most numerous; and they comprehend all those broad flat
jelly-like substances which are called by the
popular names of tangle and dulse on the
coast, and which are frequently eaten.  To this
division belong the kinds of sea-weed that are
used for making kelp; those from which iodine
is procured; those forming the celebrated
Chinese birds' nests; those sold in the oil-
shops under the name of laver; and those used
by farmers as manure.  The jointed Algæ are
very inferior in the scale of creation to the first
division; but the Confervæ (see *f* in *fig*. 149)
are well known, from the rapidity with which

they form a thick green slime, by adhering together on the surface of ditches and cisterns, and
in short, wherever there is stagnant water exposed to the open air.  The disjointed Algæ
are generally found among the Confervæ ; but
they are so small, and insignificant in appearance, as, in most cases, entirely to escape notice.

# INDEX.

---

I I

488        INDEX.

THE END.

LONDON:
BRADBURY AND EVANS, PRINTERS, WHITEFRIARS.

Printed in the United States
By Bookmasters